Mechanical Behavior of Organic Matrix Composites

Durability and Ageing of Organic Composite Materials Set

coordinated by
Marco Gigliotti

Volume 2

Mechanical Behavior of Organic Matrix Composites

Effect of Thermo-oxidative Ageing

Marco Gigliotti
Marie-Christine Lafarie-Frenot
Jean-Claude Grandidier
Matteo Minervino

iSTE

WILEY

First published 2018 in Great Britain and the United States by ISTE Ltd and John Wiley & Sons, Inc.

ISTE Ltd
27-37 St George's Road
London SW19 4EU
UK

www.iste.co.uk

John Wiley & Sons, Inc.
111 River Street
Hoboken, NJ 07030
USA

www.wiley.com

Library of Congress Control Number: 2017957134

British Library Cataloguing-in-Publication Data
A CIP record for this book is available from the British Library
ISBN 978-1-78630-018-8

Contents

List of Figures

Chapter 2

Chapter 3

Acknowledgements

The work presented in this book was partly carried out within the framework of the research project FUI COMPTINN (*composites tièdes et innovants*, 2010-2013), in collaboration with the main companies in the aeronautical field – such as AIRBUS Group Innovation (Dr Jacques Cinquin), SAFRAN (Aircraft Engines, Nacelles, Dr Isabelle Ammar-Khodja), AIRBUS SAS, SPS, ARKEMA – and several research laboratories – PIMM – Arts et Métiers ParisTech, Pprime – ISAE-ENSMA, LMT – ENS Cachan, LCTS – University of Bordeaux, LCPO – University of Bordeaux, MATEIS – INSA Lyon, IPREM – University of Pau, ICA – Armines. The authors would like to thank all those involved in this project.

They would also like to thank the members at Pprime Institute, Department of Physics and Mechanics of Materials, ENDO Group, for their support, in particular Laurence Chocinski-Arnault, David Mellier, Denis Bertheau, Médéric Morisset and Guillaume Benoit for their support with carrying out the tests.

Preface

The use of organic matrix composites (OMCs) for high-temperature structural applications is being increasingly considered, particularly in areas such as aeronautical turboengines. The integration of OMCs at these relatively high temperatures (above 100°C) is not without difficulty: subjected to mechanical loading (centrifugal forces, aerodynamic forces, etc.), they are directly exposed to environmental aggression (particularly oxygen/temperature couple).

The phenomena related to the maintenance of material in aggressive environments need to be studied in more detail, since OMCs are very sensitive to the associated degradation phenomena, such as thermo-oxidation. On the other hand, the coupling with mechanical stress cannot be overlooked since it can effectively accelerate degradation processes.

Oxidation promotes a change in the mechanical properties of the polymer matrix, leading to embrittlement, and to the development of residual strains (shrinkage), which induce the onset and the development of damage.

Previous studies have shown that matrix cracking can develop during thermal cycles in an oxidizing environment without mechanical loading being applied to the sample. At the same time, on the surface of unidirectional (UD) composites, on a more local scale, debonding at the fiber/matrix interfaces can develop spontaneously during thermo-oxidative ageing. Debonding provides new surfaces directly exposed to the environment, which, by facilitating the diffusion of oxygen, accelerate the degradation of the material and give rise to a process of self-accelerating degradation.

With the objective of predicting the onset of this damage, it is necessary to simulate strain and stress fields at different scales of the aged material. It is therefore necessary to develop models of the polymer and composite mechanical behavior depending on the level of oxidation. There are two types of challenges: the first one consists of characterizing the local nature of the phenomenon – difficult to capture by conventional mechanical tests (traction, compression, etc.) – and the other is associated with the viscous aspect of mechanical behavior, which must be identified both in the short term and in the long term.

This book is devoted to the study of the effects of thermo-oxidation on the mechanical behavior of a polymer (an epoxy resin) and carbon fiber polymer matrix composites and the characterization of the onset of damage induced by the thermo-oxidation of the matrix. This study consists of three parts:

– development of experimental methodologies, ultra-micro-indentation (UMI) and interferometric microscopy (IM), for the characterization of local mechanical behavior, depending on the thermo-oxidative ageing of the polymer, and for monitoring the matrix shrinkage associated with thermo-oxidation in composites;

– development of a local mechanical behavior law of the polymer, which is dependent on ageing, integrating inelastic strains of chemical origin, based on the thermodynamics of irreversible processes with internal variables (TIV);

– implementation, development and validation of numerical/experimental approaches by inverse analysis, for the identification of the parameters of the law of behavior of the polymer and strains of chemical origin. The identified models are used to simulate, by numerical simulation, the internal stresses associated with thermo-oxidative phenomena in composites. A simplified numerical approach, based on the Rayleigh–Ritz method, is also proposed for calculating matrix shrinkage and internal stresses in UD composites.

Marco GIGLIOTTI
Marie-Christine LAFARIE-FRENOT
Jean-Claude GRANDIDIER
Matteo MINERVINO
October 2017

Introduction

The design and manufacture of organic matrix composite (OMC) parts, in particular structural parts, is a major challenge in many industrial sectors: automotive, aeronautics, naval, etc. In aeronautical construction, there are numerous potential benefits of these materials compared with metal alloys, e.g. reduction in mass, reduction in assembly costs and losses through machining (due to the realization of complex one-shot parts), lower maintenance costs (due to excellent fatigue resistance), integration of functions, etc. Though these materials exhibit excellent specific properties under "standard" conditions, their performance degrades in the presence of aggressive environments, following the onset of ageing phenomena. This ageing often results not only from the combined action of temperature, humidity, a gaseous environment, etc., but also the in-service use conditions of the parts (type of process, curing cycle inducing residual stresses) and the structure of the material. The question then arises regarding the prediction or certification of the "lifetime" (notion of "durability") of the component under conditions of use. The lifetime can be defined as the exposure or use time beyond which a property of the material constituting the part reaches a conventionally defined threshold.

In this book, only composites made of carbon fibers and thermosetting matrices that constitute aircraft parts are studied, in the context of their use in “warm” zones, e.g. for parts close to engines.

> “The engine cowls of a helicopter are made of composite parts; the temperatures are approximately 150°C, continuously. Epoxy matrices can be used, but the relatively high operating temperature for this type of material causes accelerated ageing of the structure by accelerating cracking phenomena, which shortens the life of the part.” (translated by [CIN 02])

These composites can be considered as materials/structures, in which only the polymer matrix and possibly the fiber/matrix interfaces are affected by ageing. Therefore, in order to be able to predict or certify the durability of these composites as a function of the ageing of their constituent materials, it makes sense to refer first of all to work on the ageing of polymers.

According to Fayolle and Verdu [FAY 05], ageing is any phenomenon of slow and irreversible evolution of the microstructure and/or composition of a material under the effect of its own instability, interaction with the environment, mechanical stress or the combination of several of these causes (known as couplings). Any phenomena that cannot be understood in a time scale compatible with technological and economic constraints, and that therefore requires the use of accelerated ageing tests and a model allowing extrapolation of the results of the latter are called “slow” phenomena [RIC 11]. Two categories of ageing can be distinguished: physical ageing and chemical ageing.

Physical ageing encompasses all ageing processes in which the chemical structure of the macromolecules is preserved. It includes structural relaxation processes (increase in compactness, modification of the conformation of the chains), slow crystallization and all processes involving mass transfer (absorption–desorption phenomena) [FAY 05]. There has been a lot of research work into this type of ageing – both on phenomenological aspects and on their modeling – ageing which is of the first order, when polymers are used at temperatures that are lower but close to their glass transition.

Chemical ageing is an evolution of the chemical structure under the influence of the environment [VER 13]. Users of polymer materials classify the different types of chemical ageing according to what is supposed to be the main degradation factor: photochemical, radiolytic, hydrolytic, thermo-oxidative, biochemical, etc. The light intensity, the dose rate of the ionizing radiation, the concentration of water, oxygen, etc., are the influential factors. In the temperature domain where ageing mechanisms are (sometimes deemed) identical, temperature only acts on the reaction kinetics, which allows the proposal of "accelerated ageing" tests.

The difficulties associated with the experimental characterization of ageing phenomena (type of test, type of conditioning, type of test piece), the determination of reliable acceleration conditions and the "measurement" of the lifetimes of composites with an organic matrix contribute to the complexity of research in this field. European (JAA) or American (FAA) certification bodies have imposed standards for dimensioning Organic Matrix Composite (OMC) structures to take into account environmental conditions and set the conditions for accelerated ageing testing (standard JAR 25 – 603, [MOR 05]): most of these standards are pragmatic, based on empirical approaches and validated by feedback.

A scientific approach to predicting the lifetime of OMC materials, subject to an aggressive environment leading to ageing, requires:

– a thorough understanding of degradation mechanisms;

– their modeling (functions of time and temperature, even of space when diffusion phenomena are to be taken into account);

– the establishment of procedures for identifying the parameters of these models.

If all these steps are validated, the "long-term" extrapolation of these models becomes possible.

As mentioned, one of the most aggressive ageing processes for organic matrix composite materials is chemical ageing, particularly thermo-oxidative ageing. The latter results in the onset of matrix "shrinkage", possibly associated with fiber/matrix decohesions, observable on the surfaces exposed to the oxidizing environment [LAF 04].

To study the phenomena associated with this type of ageing, a specific experimental protocol has been developed (more details will be given in Chapter 1 of this book).

This protocol relies essentially on the observation and measurement, using an optical profilometer based on interferometric microscopy (IM), of the reliefs of one of the external surfaces of carbon/epoxy composite samples (see Figure I.1). In this study, the material under consideration consists of continuous HTS fibers and the TACTIX matrix. Samples that are 15 × 15 mm in size were cut into a 10 mm (50 unidirectional ply) plate and the observations were made on a surface orthogonal to the direction of the fibers. The test comprises the following steps (Figure I.1):

– polishing the surface to be observed, according to an optimized protocol;

– observation of the polished surface, in the initial state (virgin);

– conditioning under a set temperature and possibly pressure of the surrounding gas environment;

– observation of the surface after ageing.

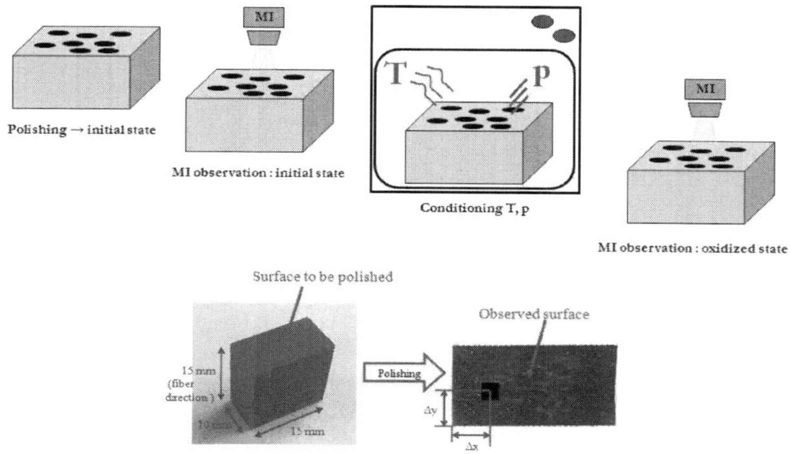

Figure I.1. *Observation by interferometric microscopy (IM) of the surfaces of aged UD samples. For a color version of this figure, see www.iste.co.uk/gigliotti/mechanical.zip*

The optical profilometer used (Talysurf CCI 6 000, Taylor & Hobson®) is shown in Figure I.2. Its principle is based on interferometry:

– a light beam is separated into two parts, one of which is directed to the surface of the sample and the other to a reference mirror. The recombination of the reflected beams, when the sample is at the focal distance, gives a figure of interference fringes;

– by positioning the focal plane of the lens at different depths in the sample and by eliminating all light from the points out of focus, it is possible to produce a series of images from which a three-dimensional representation of the object can be obtained;

– post-processing software reconstructs a digital image of the surface.

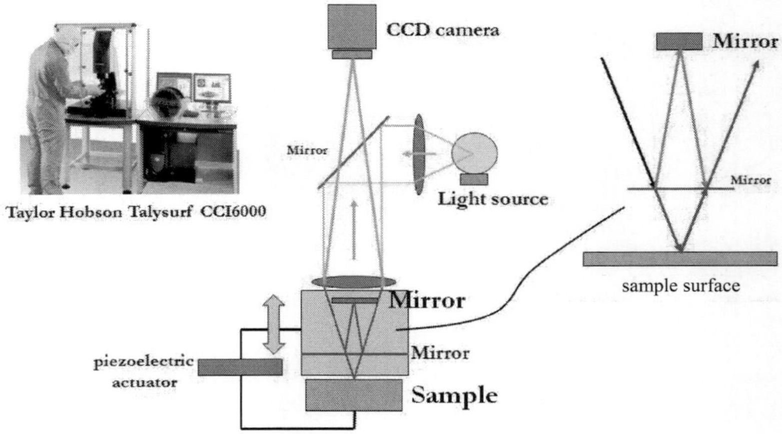

Figure I.2. *The principle of IM. For a color version of this figure, see www.iste.co.uk/gigliotti/mechanical.zip*

For this device, the vertical displacement of the objective is imposed by a piezoelectric actuator, which allows a high vertical resolution to be obtained (0.01 nm, depending on the manufacturer). On the other hand, the measuring range and the spatial resolution depend on the magnification of the lens. In the rest of the study, for our measurements, a magnification of 50 × is systematically used, allowing the observation of a surface area of 360 × 360 μm^2, with a spatial resolution of approximately 0.35 μm.

With the chosen optics, it is not possible to measure a surface or area of the surface with a slope greater than 27.7° with respect to the horizontal plane. This limit of the apparatus was exploited by Vu *et al.* [VU 12] for the study of matrix shrinkage profiles on UD composite surfaces. In this case, the authors have demonstrated, by comparison with images obtained by scanning electron microscopy (SEM), that the decohesions at the fiber/matrix interfaces result in "unmeasured" points on the surface.

An example of 3D reconstructions of a part (25 × 25 µm) of the surface of a virgin and aged UD sample is given in Figure I.3. In this image, the fibers of the matrix between fibers are clearly distinguished. In the initial state, the matrix shrinkage between fibers is relatively small, the maximum value being approximately 250 nm; this shrinkage may be associated with the cooling phase, which takes place after the curing of the composite, and the relatively high coefficient of thermal expansion (of approximately 50×10^{6} 1/°C) of the polymer matrix. In the oxidized state, i.e. in Figure I.3, 72 hours (3 days) under 2 bar oxygen at 150°C, the increase in the matrix shrinkage is clearly observed, which increases from approximately 150 nm to 1.5 µm, and the presence of debonding along the perimeter of the fibers.

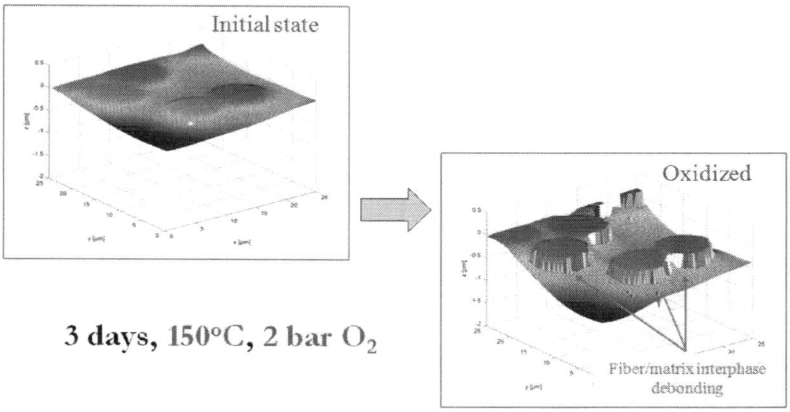

Figure I.3. *3D reconstruction of the surface (25 × 25 µm) of UD samples (virgin and aged) observed by IM (see also [GIG 16b, GIG 16c]). For a color version of this figure, see www.iste.co.uk/gigliotti/mechanical.zip*

The interpretation of these observations is as follows: following the exposure to high temperatures (possibly under high partial pressure of oxygen), the carbon fibers being inert to temperatures around 300°C, the O_2 penetrates into the composite matrix from the exposed surfaces towards the core of the sample, with complex diffusion/reaction kinetics. These kinetics have been specified in detail in [COL 12, VER 13], using a mechanistic schematic that describes in detail the chronology of the main chemical reactions occurring during thermo-oxidation. The O_2 diffusion/reaction generates O_2 concentration gradients, accompanied by gradients of mechanical matrix properties (stiffness, toughness, etc.) and chemical strains.

The gradients of chemical origin are directly related to the variations in mass and density gradients associated with the generation of volatile reaction products [DEC 03, COL 12], which may result in changes in the shape of pure polymer samples or matrix shrinkage in composites. The creation of reaction volatiles also produces considerable mass losses in the polymer and composite samples [COL 12]. The mechanical property gradients are associated with macromolecular chain scission promoted by the oxidation reaction, which in turn are responsible for the embrittlement of the macromolecular network and the changes in its time-dependent behavior ("antiplastification" phenomenon [VER 13]).

The objective of this book is to specify the effects of thermo-oxidative ageing on the mechanical behavior of carbon fiber composites and thermosetting matrices, using an experimental/numerical approach based on the use of dedicated experimental techniques, and in the context of the thermodynamics of irreversible processes with internal variables (TIV). The phenomenon of embrittlement will not be explicitly dealt with: attention will be paid to local changes in mechanical behavior, characterized by ultra-micro-indentation (UMI) tests and chemical inelastic strains measured by interferometric microscopy (IM) on unidirectional (UD) composite samples.

The book is organized as follows:

– Chapter 1 presents the phenomenological aspects of thermo-oxidative ageing of OMCs, with an overview of ultra-micro-indentation tests for the characterization of the local mechanical behavior of virgin and aged resins by thermo-oxidation, and the use of interferometric microscopy for the characterization of matrix shrinkage induced by thermo-oxidation in UD composites.

– Chapter 2 presents the modeling of thermo-oxidative ageing of composites, with an introduction to the thermodynamics of irreversible processes with internal variables (TIV), the development of a law of behavior of the polymer resin dependent on ageing and chemical inelastic strains.

– Chapter 3 presents model identification, numerical simulations of composites at the microscopic scale, and validations by comparison with experimental measurements.

1

Phenomenological Aspects of Thermo-oxidative Ageing of OMCs

The studied material is a "high-performance" composite for aerospace applications, the carbon/epoxy HTS/TACTIX composite, supplied by Airbus Group Innovations (AGI). The matrix of this composite is an epoxy/amine polymer consisting of a mixture of three components: tris(4-hydroxyphenyl)methane triglycidyl ether (TACTIX 742), bisphenol A diglycidyl ether (TACTIX 123 or DER 332) and 4,4'-dichlorodiphenyl sulfone (HT976 or DDS), an amine with a crosslinking function. The glass transition temperature (T_g) of the resin (supplied by the manufacturer and obtained by *Dynamic Mechanical Analysis* (DMA)) is between 190°C and 250°C. The mass ratio of resin in the composite (pre-preg) is approximately 34%. The polymerization cycle of the composite consists of a gel phase of 1 hour at 140°C, an initial dwell of 3 hours at 180°C and a second dwell of 1 hour at 250°C. The material is then subjected to post-curing under vacuum for 10 hours at 230°C, essential to obtain a degree of crosslinking of approximately 100%.

There are two possible ways to accelerate the thermo-oxidation phenomena affecting the organic resins: they involve increasing either the ageing temperature or the oxygen pressure. On the one hand, increasing the ageing temperature reduces the activation energy of the chemical reactions and, on the other hand, the properties of the resin are greatly modified. The effectiveness of pressure as a thermo-oxidation acceleration parameter has been studied by several authors (see, among others, [GRA 15] and the cited references): in particular in Vu *et al.* [VU 12], it has been demonstrated that "moderate" (2 bar) pressure values are suitable for accelerating thermo-oxidation phenomena without inducing other parasitic degradation mechanisms.

Figure 1.1 shows the "COMEDI" test setup, made or adapted to the laboratory for the ageing of materials in a controlled environment. The main technical characteristics of this setup are as follows:

– maximum temperature: 200°C;

– maximum pressure: 10 bar;

– environment: air, oxygen, nitrogen.

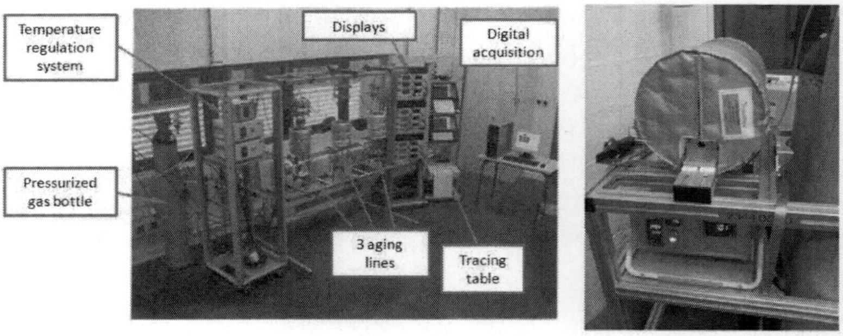

Figure 1.1. *"COMEDI" test setup for material ageing (see also [GIG 16a, MIN 13, MIN 14]). For a color version of this figure, see www.iste.co.uk/gigliotti/mechanical.zip*

The most obvious effects of degradation of a polymer by thermo-oxidation are undoubtedly the change in color (the sample changes from a light color – normally yellow for an epoxy resin (see Figure 1.2(a)) to a darker color (brown or black) in the aged state (see Figure 1.2(b))) and the change in mass of the sample. In particular, mass change is very important; it is easily measured by simple gravimetric tests and it is usually the key ingredient for the identification of reaction/diffusion model parameters [COL 12].

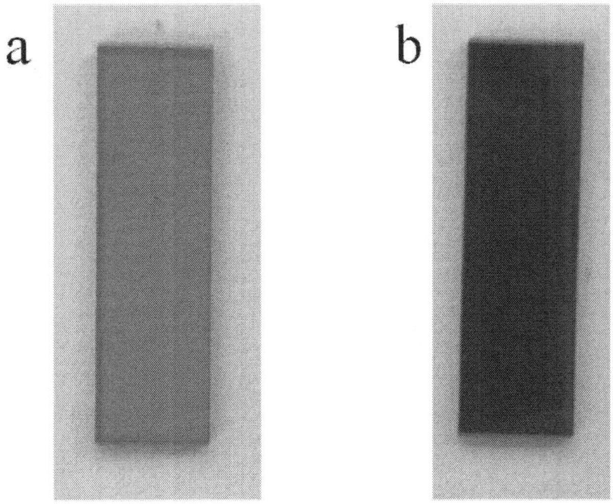

Figure 1.2. *Color change of TACTIX resin samples induced by thermo-oxidation (see also [GIG 16a, MIN 13, MIN 14]). For a color version of this figure, see www.iste.co.uk/gigliotti/mechanical.zip*

Figure 1.3 shows the mass evolution of the TACTIX resin samples aged in air at atmospheric pressure and under 2 bar oxygen at a temperature of 150°C. This temperature corresponds to the average values of the use temperatures foreseen for this polymer.

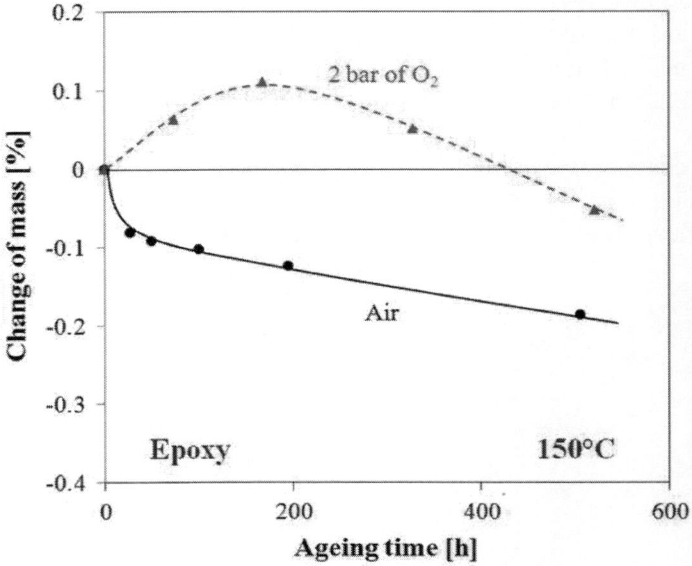

Figure 1.3. *Change in mass of TACTIX resin samples aged in air at atmospheric pressure and under 2 bar O_2 at a temperature of 150°C (see also [GIG 16a, MIN 13, MIN 14]). For a color version of this figure, see www.iste.co.uk/gigliotti/mechanical.zip*

We note that:

– the oxidized samples under atmospheric air show a slow, but progressive, decrease in mass during the initial hours of ageing;

– for samples oxidized under 2 bar oxygen, there is an initial increase in mass, which reaches a peak around 160 hours of ageing, followed by a decrease in mass.

In fact, the mass changes due to the competition between the diffusion of oxygen, bringing new molecules into the chemical structure, and the chemical oxidation reactions, leading to the formation of volatile products and water molecules, which rapidly diffuse outwardly from the sample, resulting in a decrease in mass.

The competition between intake and loss of mass is affected by diffusion and reaction kinetics, which depend on the nature of the polymer or the ageing conditions (temperature, pressure, etc.) or the local oxidation rate of the material.

The thermomechanical behavior of virgin and aged resin samples is initially characterized by DMA tests, by measuring a "complex" modulus composed of an elastic part (E′, conservation modulus), corresponding to the phasing response, and a viscous part (E″, loss modulus), corresponding to the dephasing response.

Figure 1.4 shows the DMA spectra of the TACTIX resin in the initial state between –150°C and 300°C: the conservation modulus E′ (see Figure 1.4(a)) and the loss modulus E″ (see Figure 1.4(b)).

At low temperatures, the modulus E″ peaks at about –75°C; at the same time, the conservation modulus decreases from 4,600 MPa (at –100°C) to 3,500 MPa (at –50°C). This indicates that a secondary relaxation phenomenon takes place at these low temperatures, corresponding to the so-called 'β' transition. At high temperatures, the DMA spectra do not have a conventional appearance for an epoxy resin. Several relaxations, which are difficult to identify, are simultaneously active, leading to a drop in E′ and a "peak" in E″, over a fairly wide temperature range. Therefore, the glass transition (T_g) of the TACTIX resin cannot be accurately identified, but it is within a temperature range between 200°C and 250°C.

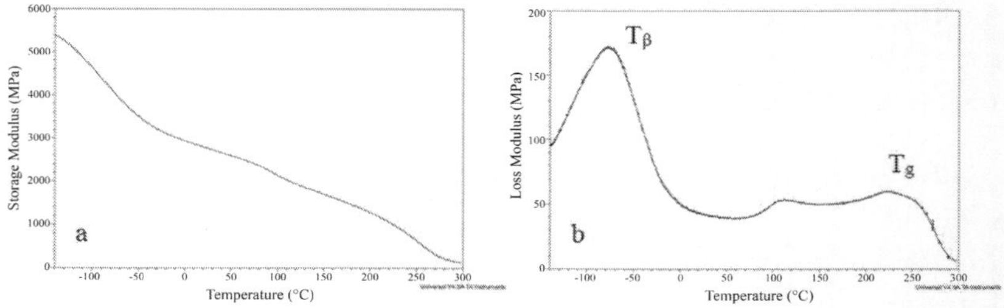

Figure 1.4. *DMA spectra of the TACTIX resin in the initial state: a) conservation modulus E' and b) loss modulus E" (see also [GIG 16a, MIN 13, MIN 14])*

Figure 1.5 illustrates the DMA spectra – conservation modulus (Figure 1.5(a)) and loss modulus (Figure 1.5(b)) – of TACTIX resin samples oxidized at 150°C under 2 bar O_2 for different durations of up to 5 days of ageing.

At room temperature, the E′ modulus increases from 2,800 MPa in the virgin state to 3,440 MPa after 5 days of ageing: no significant change in E′ is observed at low temperature (–140°C). Meanwhile, the intensity of the β transition decreases as the oxidation increases (see Figure 1.5(b)), which explains the variation in the observed conservation modulus: a smaller variation in the modulus E′ with temperature at a lower β transition peak amplitude, which leads to a higher value at room temperature. This phenomenon has been observed and studied by several authors and is known as the antiplasticization phenomenon [VER 13]. Moreover, according to Figure 1.5(b), the peak associated with the glass transition appears to be better defined when the oxidation time increases, and the glass transition temperature decreases to 143°C from 3 days of ageing under 2 bar O_2.

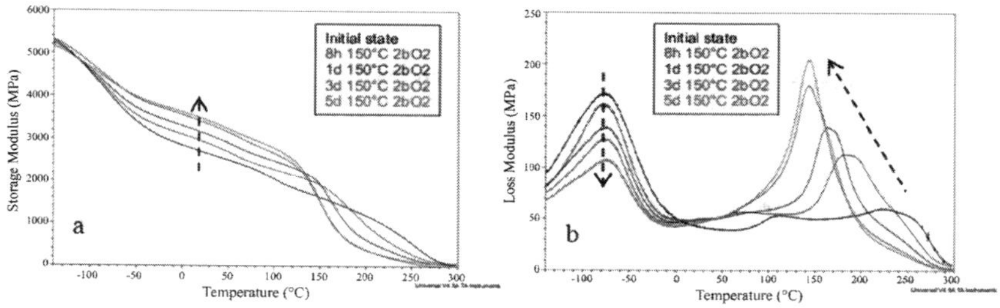

Figure 1.5. *DMA spectra of TACTIX resin samples oxidized at 150°C under 2 bar O_2 for different durations of up to 5 days of ageing: a) conservation modulus and b) loss modulus (see also [GIG 16a, MIN 13, MIN 14]). For a color version of this figure, see www.iste.co.uk/gigliotti/ mechanical.zip*

1.1. Effect of thermo-oxidation on the local mechanical behavior of the polymer

Figures 1.6 and 1.7 schematically show the protocol for the preparation of pure, virgin or oxidized resin samples for ultra-micro-indentation (UMI) tests, respectively.

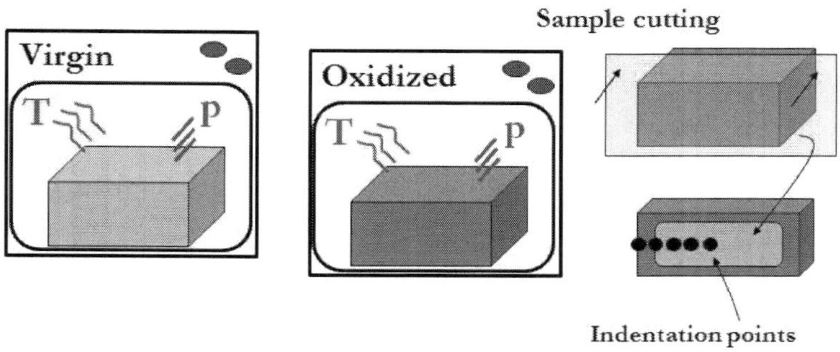

Figure 1.6. *Schematic representation of ultra-micro-indentation (UMI) tests. For a color version of this figure, see www.iste.co.uk/gigliotti/mechanical.zip*

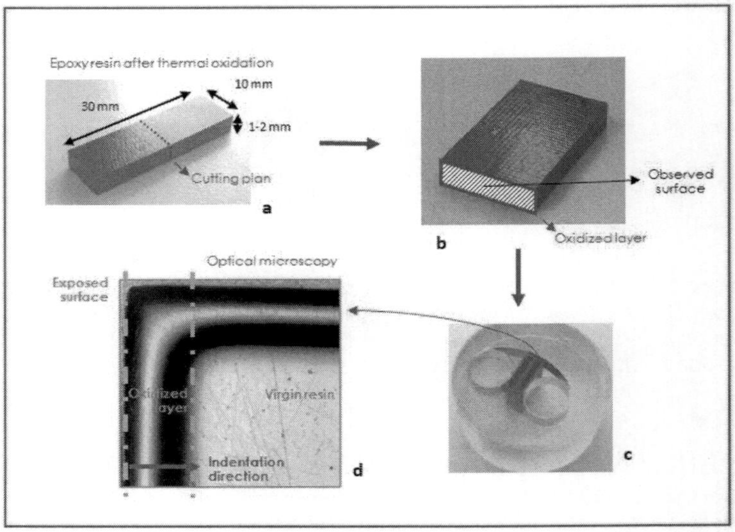

Figure 1.7. *Protocol for preparing samples for UMI tests (see also [GIG 16a, MIN 13, MIN 14]). For a color version of this figure, see www.iste.co.uk/gigliotti/mechanical.zip*

The aged samples are cut in the center in a direction perpendicular to the surface directly exposed to the environment (see Figures 1.6 and 1.7(a)). This allows access to the internal part of the sample, which is less affected by oxidation phenomena. Figure 1.7(b) shows the observed surface, or indentation surface: an oxidized layer, in which the properties of the polymer vary, is present at its edges. The two parts of the sample are then placed side by side, coated and, finally, polished by means of a semi-automatic polishing machine to obtain a flat indentation surface. The polishing of the surface involves different steps, using 3 μm and 1 μm polishing papers and a special lubricant, with a very low percentage of alcohol, to avoid surface degradation. The final state of the sample is shown in Figure 1.7(c). Figure 1.7(d) shows an optical microscopic image of the surface of an aged TACTIX resin sample. The oxidized layer can be easily distinguished from the non-oxidized polymer core, e.g. in Figure 1.7(d), we can see that the TACTIX resin sample, aged for 72 hours at 150°C under 2 bar O_2, has an oxidized layer with a thickness of 150 μm.

The ultra-micro-indentation apparatus available in the laboratory is Fischerscope® H100C (see Figure 1.8(a)), equipped with a square-based pyramidal Vickers diamond tip, and the actual shape (including small imperfections and surface defects) is known thanks to 3D mapping, which is obtained by three-dimensional profilometry.

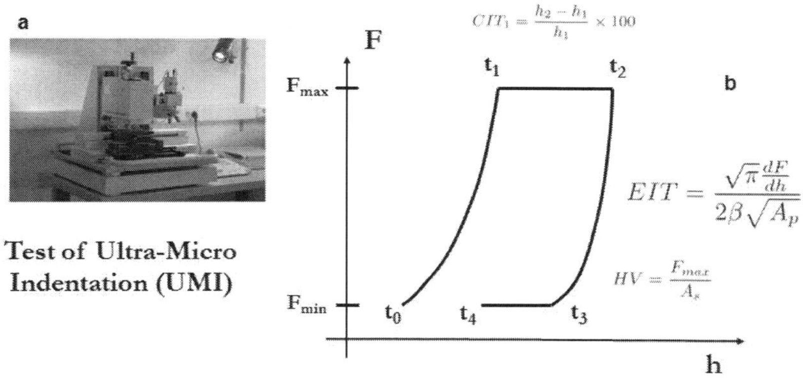

Figure 1.8. *a) Fischerscope® H100C for ultra-micro-indentation (UMI) tests; b) schematic of a load-indentation depth curve by UMI (see also [GIG 16a, MIN 13, MIN 14]). For a color version of this figure, see www.iste.co.uk/gigliotti/mechanical.zip*

The micro-indentation test allows localizing the stress in a small volume. For the studied polymers, a load of 5 mN leads to an indentation depth of approximately 1 μm. With this level of penetration, the square indentation print has a diagonal of less than 10 μm. Moreover, the localization of the stresses generates larger strains – by approximately 40% to 50% (see e.g. [TVE 11]) – and large strain gradients, localized in a few micrometers (~5 μm from the diamond tip); the values are not accessible with conventional tensile or compression tests. During the indentation test, the load F and the indentation depth h are recorded and allow plotting a "load versus indentation depth" curve (see Figure 1.8(b)): this curve usually consists of three phases: a loading phase ($t_0 < t < t_1$), a creep phase ($t_1 < t < t_2$) and an unloading phase ($t_2 < t < t_3$). The parameters of the three phases (creep duration, loading speed) can be modified to allow the polymer behavior to be studied under different loads and with different test protocols. The indentation curve can be interpreted according to the method of Oliver and Pharr [OLI 04], through measurements of the elastic modulus of indentation (EIT) and the hardness (HV). The EIT module is calculated directly from the indentation curve using equation [1.1]:

$$\mathrm{EIT} = \frac{\sqrt{\pi}\,\mathrm{dF/dh}}{2\mathrm{b}\sqrt{\mathrm{A_p}}} \qquad [1.1]$$

where b is related to the indenter shape (b = 1.013 for a Vickers), A_p is the contact area between the diamond tip and the sample, projected on a plane perpendicular to the indentation axis, and dF/dh is the slope of the indentation curve at the beginning of the unloading phase.

1.1.1. *Oxidized layers*

Figure 1.9 shows images obtained by optical microscopy of the indentation surfaces of four oxidized TACTIX resin samples: after 24 hours (see Figure 1.9(a)), 48 hours (Figure 1.9(b)), 72 hours (Figure 1.9(c)) and 120 hours (Figure 1.9(d)) of ageing at 150°C under 2 bar O_2. All images were taken using a 20 x lens, which allows for the direct comparison of different dimensions of oxidized layers.

According to [COL 05], the depth of the oxidized layers, which can be observed as the dark parts on the images, increases during conditioning, with the thickness increasing from 90 μm after 24 hours of ageing (see Figure 1.9(a)) to 230 μm after 120 hours (see Figure 1.9(d)).

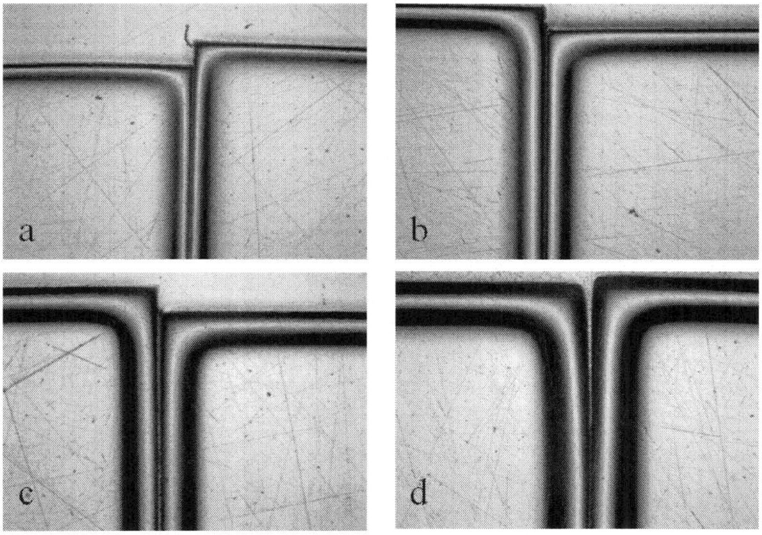

Figure 1.9. *Images by optical microscopy (OM) of indentation surfaces of four TACTIX resin samples oxidized at 150°C under 2 bar O_2: a) 24 hours; b) 48 hours; c) 72 hours and d) 120 hours (see also [GIG 16a, MIN 13, MIN 14])*

In order to better understand why we observe variations in contrast on the surface of aged samples using optical contrast microscopy (OM), the same surfaces were observed using interferometric microscopy (IM). Figure 1.10(a) identifies the location on the surface, which is selected and observed by interferometric microscopy, and Figure 1.10(b) shows the result of the measurement.

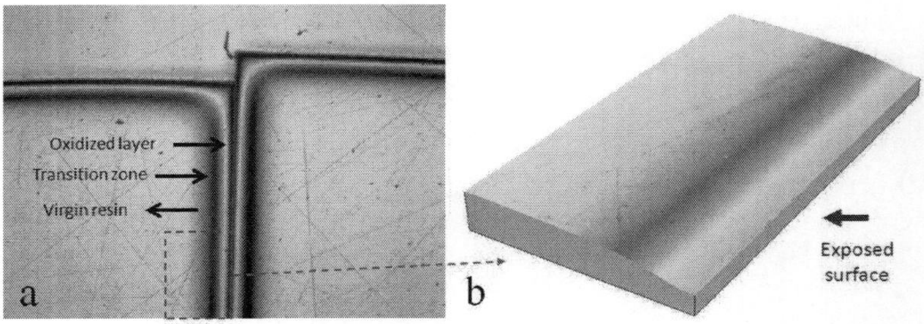

Figure 1.10. *a) Image using OM and b) IM of the indentation surface of a TACTIX resin sample oxidized for 24 hours at 150°C under 2 bar O_2 (see also [GIG 16a, MIN 13, MIN 14]). For a color version of this figure, see www.iste.co.uk/gigliotti/mechanical.zip*

In Figure 1.10(b), the colors are associated with variations in depth with respect to a reference surface, changing from pink to green when oxidation increases. These measurements reveal material shrinkage within the oxidized layer, which increases as we get closer to the surface directly exposed to the environment. These differences in height are responsible for the color contrast observed by optical microscopy.

Figure 1.11 shows the evolution of shrinkage during oxidation: in the initial state (see Figure 1.11(a)), the indentation surface is flat and well polished, while the oxidized states (Figure 1.11(b), (c), (d), (e)) display shrinkage, which increases as the oxidation time increases.

From the 3D images in Figure 1.11, the vertical displacements of the surface can be measured and plotted as a function of the distance to the surface exposed to the environment, and the evolution of these profiles can be characterized as a function of oxidation time (see Figure 1.12). We therefore observe that the maximum shrinkage measured at the edge of the sample greatly increases with the ageing time: from approximately 100 nm in the initial state, reaching approximately 2 μm after

24 hours of ageing and 4 μm after 120 hours at 150°C under 2 bar oxygen.

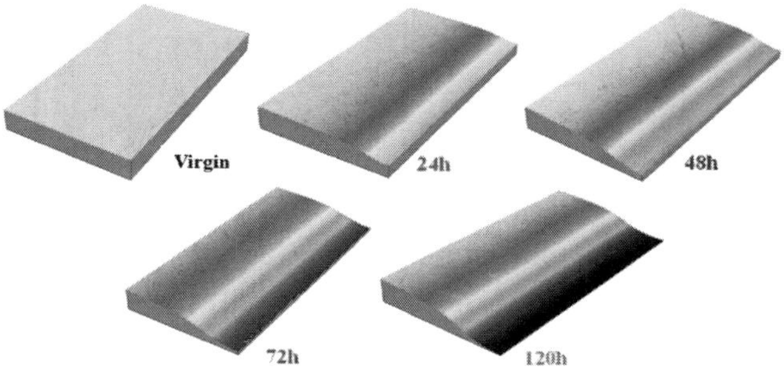

Figure 1.11. *Images using IM of indentation surfaces of TACTIX resin samples: a) in the initial state and for four oxidized samples; b) at 24 hours; c) at 48 hours; d) at 72 hours and e) at 120 hours at 150°C under 2 bar O_2 (see also [GIG 16a, MIN 13, MIN 14]). For a color version of this figure, see www.iste.co.uk/gigliotti/mechanical.zip*

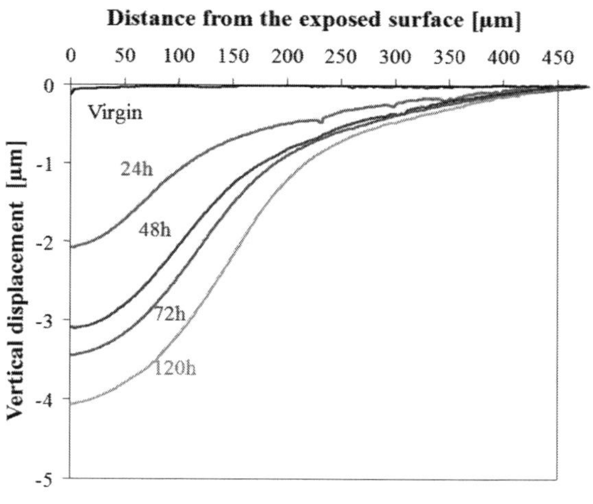

Figure 1.12. *Vertical displacements as a function of distance from the surface exposed to the environment using IM on indentation surfaces of TACTIX resin samples: in the initial state and for four oxidized samples, 24 hours, 48 hours, 72 hours and 120 hours at 150°C under 2 bar O_2 (see also [GIG 16a, MIN 13, MIN 14]). For a color version of this figure, see www.iste.co.uk/gigliotti/ mechanical.zip*

The steady increase in shrinkage, as well as the profiles obtained as a function of the distance to the surface in contact with the environment (shown in Figures 1.11 and 1.12), suggests a link between surface shrinkage and oxidation gradients. During thermo-oxidation, residual shrinkage strains of chemical origin develop in the oxidized layer. At the time the sample is cut and polished, these strains relax, producing out-of-plane surface displacements. The measurement of the shrinkage of the oxidized layer by IM allows confirmation of the existence of a residual strain gradient of chemical origin, which develops from the surface exposed to the environment towards the core of the polymer and suggests a method for its characterization. It should be noted that, for all the tested samples, the intensity of the matrix shrinkage has no impact on the UMI measurements, with the slope of the tested surfaces being significantly less than 1%.

1.1.2. *EIT measurements by UMI*

The indentation tests were carried out using the following protocol: a loading phase with a velocity of 0.25 mN/s up to 5 mN, and an unloading phase at the same speed, without any holding phase. The measurements are taken at different distances from the surface exposed to the environment, in order to demonstrate the gradient of properties which appears during oxidation: for each indentation print, an indentation curve and an EIT modulus can be measured; at each distance from the surface exposed to the environment, eight indentation prints are made, making it possible to identify an average indentation curve and to calculate an average indentation modulus. Figure 1.13(a) shows preliminary EIT module measurements taken previously on the surface of a sample of virgin TACTIX resin at randomly chosen points: the obtained modulus values are homogeneous over the entire surface of the sample and vary between 3,350 MPa and 3,450 MPa. Moreover, Figure 1.13(b) illustrates EIT "profiles" obtained on a sample maintained at 150°C for 48 hours under 2 bar nitrogen (N_2), and measured at different distances from the surface exposed to the environment. In this inert environment, the indentation modulus of the polymer remains unchanged and uniform in the thickness of the sample.

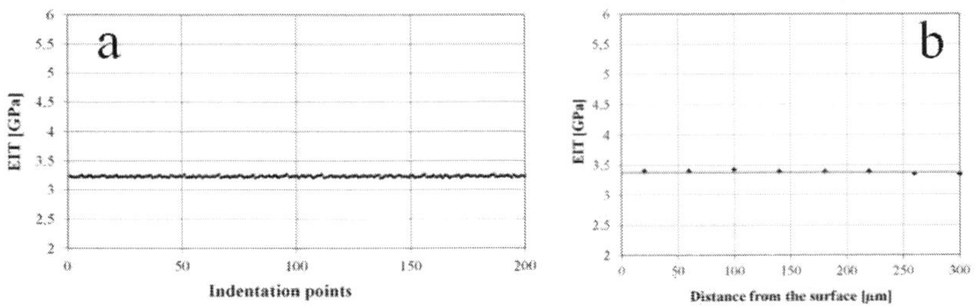

Figure 1.13. *EIT measurements for TACTIX resin: a) virgin and b) after 48 hours remaining in a neutral environment, 2 bar N_2, at 150°C (see also [GIG 16a, MIN 13, MIN 14])*

Figure 1.14 shows the EIT profiles measured on samples aged in air at atmospheric pressure (up to 400 hours, see Figure 1.14(a)) and 2 bar O_2 (up to 330 hours, Figure 1.14(b)), at 150°C: at a distance of 40 μm from the surface exposed to the environment, after 400 hours in air and 330 hours under 2 bar O_2, the EIT modulus varies by approximately +62% and +80% of the initial value, respectively.

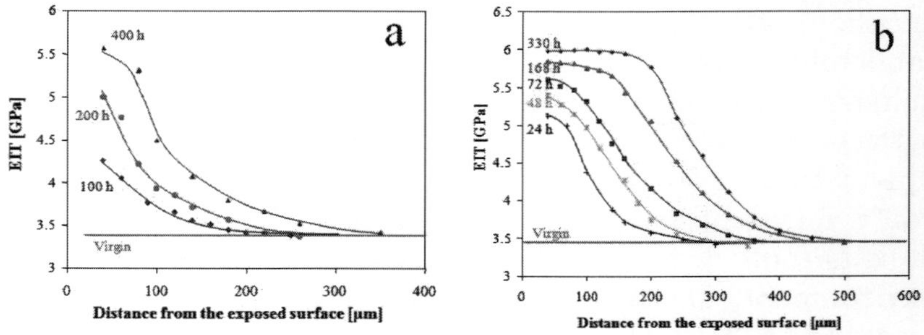

Figure 1.14. *Profiles of the EIT modulus, for TACTIX resin: a) after ageing in air and b) under 2 bar O_2, at 150°C (see also [GIG 16a, MIN 13, MIN 14]). For a color version of this figure, see www.iste.co.uk/gigliotti/mechanical.zip*

It can be seen that the indentation profiles strongly depend on the type of coating: the thickness of the zones affected by the oxidation (where changes in the EIT values compared with the initial values can be appreciated) is markedly different in the two cases. Studies in the literature [OLI 08] have shown, for materials similar to that of this study, that there is a correlation between the EIT profiles and the concentrations gradients of the oxidation products, i.e.

$$\mathrm{EIT}\,(\mathrm{Q}) = \mathrm{A}(1 - \exp^{-\mathrm{BQ}}) \quad [1.2]$$

where A and B are constants and Q is the concentration of oxidation products (see [COL 12] and [OLI 08] for a precise definition of the Q parameter and a detailed description of the chemical phenomena involved in the calculation of this parameter). This is an extremely significant result, as it is possible to state that the measurements of the EIT modulus can be compared, albeit from a purely phenomenological point of view, with the physicochemical mechanisms of thermo-oxidation. The relative variation in the EIT modulus values (measured at a distance of 40 μm from the surface exposed to the environment) and the increase in the thickness of the oxidized layer (defined as the thickness where the value of the EIT modulus is greater than the initial modulus by more than 4%) are plotted as a function of the oxidation time in Figures 1.15(a) and 1.15(b), respectively. The shape of these curves (acceleration and then stabilization of behavior) qualitatively reflects the competition between oxygen diffusion and reaction within the material [COL 12].

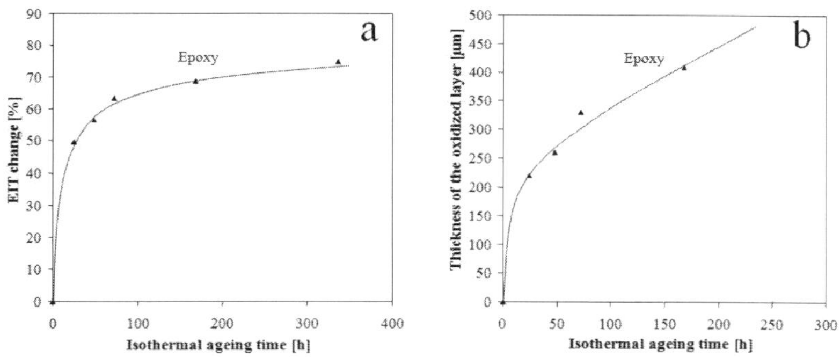

Figure 1.15. *Variation: a) in the EIT modulus and b) in the thickness of the oxidized layers as a function of the ageing time at 150°C under 2 bar O_2 (TACTIX resin) (see also [GIG 16a, MIN 13, MIN 14])*

In Figure 1.16, the indentation profile of Figure 1.14(b) is superimposed onto the image obtained by optical microscopy in Figure 1.9, for a sample aged for 120 hours under 2 bar O_2 at 150°C: this superimposition reveals a link between the EIT modulus, the thickness of the layer affected by thermo-oxidation and shrinkage of chemical origin occurring within this layer. However, it can be seen that the variations in EIT extend beyond the dark layer, which seems to show that measuring the thickness of the oxidized layer using optical microscopy gives underestimated values.

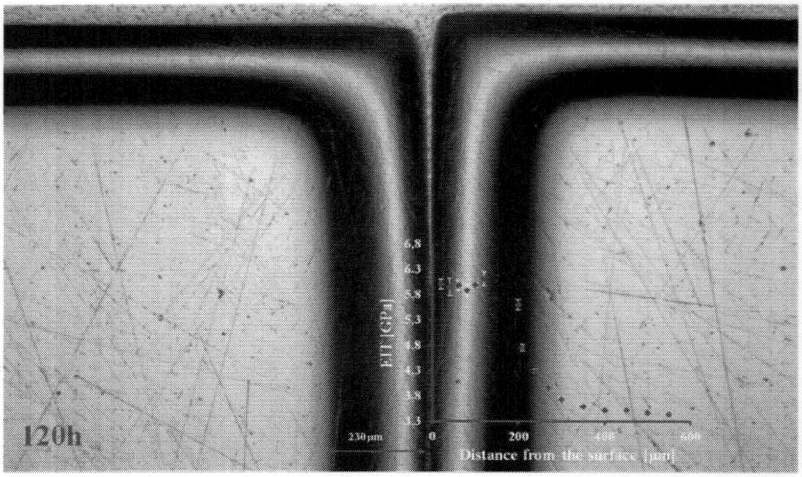

Figure 1.16. *EIT profiles obtained for TACTIX resin aged under 2 bar O_2 at 150°C (120 hours) superimposed on OM images of oxidized surfaces (see also [GIG 16a, MIN 13, MIN 14])*

The cross-sectional analysis of Figure 1.14(a) and 1.14(b) also reveals the effect of the partial pressure of oxygen on the EIT profiles: in Figure 1.17, two EIT modulus profiles obtained for two samples aged at 150°C are compared, one (blue in the figure) kept for 200 hours in air at atmospheric pressure, and the other (red) under 2 bar O_2 for 24 hours. For these two samples, the maximum indentation moduli, measured at 40 µm from the surface exposed to the environment, are approximately 5,100 MPa, thus showing the accelerating role of the oxygen pressure. However, the shapes of the profiles are not exactly the same: the sample aged in air shows a faster variation of the modulus in the thickness, resulting in a thinner oxidized layer of approximately 100 µm. The slightly different shapes of the profiles can easily be explained by a faster oxygen propagation speed along the depth as the oxygen pressure increases.

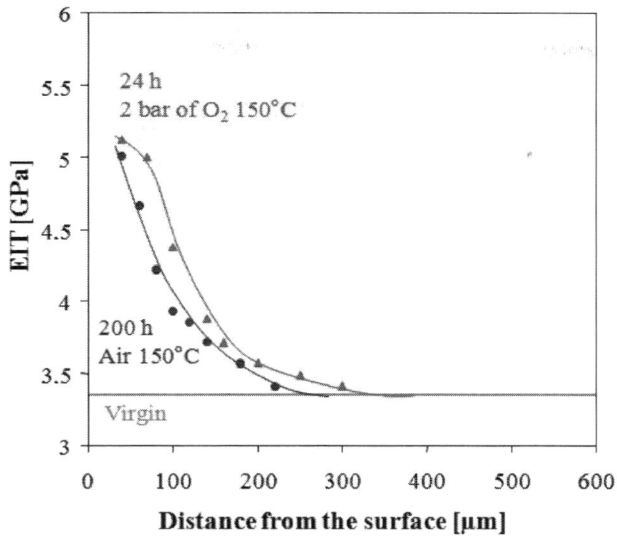

Figure 1.17. *Comparison of EIT modulus profiles, for TACTIX resin, after ageing at 150°C for 24 hours under 2 bar O_2 (red curve) and 200 hours in air at atmospheric pressure (blue curve) (see also [GIG 16a, MIN 13, MIN 14]). For a color version of this figure, see www.iste.co.uk/gigliotti/mechanical.zip*

1.1.3. *Introduction to the parameter γ, an oxidation tracer*

Based on the existence of experimental correlations between the EIT modulus values and the oxidation product concentration, Q, in this study, we chose to replace the parameter Q with the "dimensionless EIT" modulus, which we will call the "phenomenological ageing parameter (γ)". This parameter allows accounting for the local level of oxidation in the thickness of the sample. The definition of γ is as follows:

$$\gamma(T, p, t, z) = \frac{EIT}{EIT_0} - 1 \qquad [1.3]$$

where T is the ageing temperature, p is the oxygen partial pressure, t is the oxidation time, z is the distance to the surface exposed to the environment and EIT_0 is the indentation modulus of the unoxidized virgin resin. For a given temperature and partial pressure of oxygen, profiles of γ as a function of z and t can then be plotted. For example, Figure 1.18 shows the profiles of γ obtained in air at atmospheric pressure (a) and under 2 bar O_2 (b), at 150°C. The value of γ outside the oxidized layer is zero; in this layer, this parameter is strictly positive and reaches, e.g. a value of 0.74 after 330 hours of oxidation under 2 bar O_2.

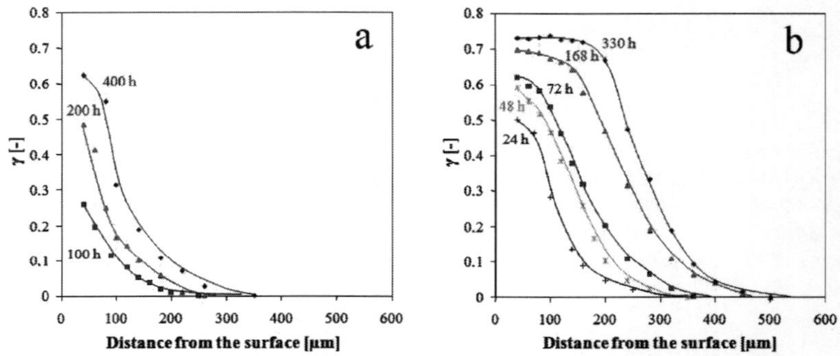

Figure 1.18. *Profiles of the parameter γ obtained for the TACTIX resin aged at 150°C: a) in air at atmospheric pressure and b) under 2 bar O_2 (see also [GIG 16a, MIN 13, MIN 14]). For a color version of this figure, see www.iste.co.uk/gigliotti/mechanical.zip*

Figure 1.19 shows the evolution of γ_{max} (a value of γ taken as close as possible to the sample surface), as a function of the ageing time in air (red curve) or under 2 bar O_2 (black curve), at 150°C.

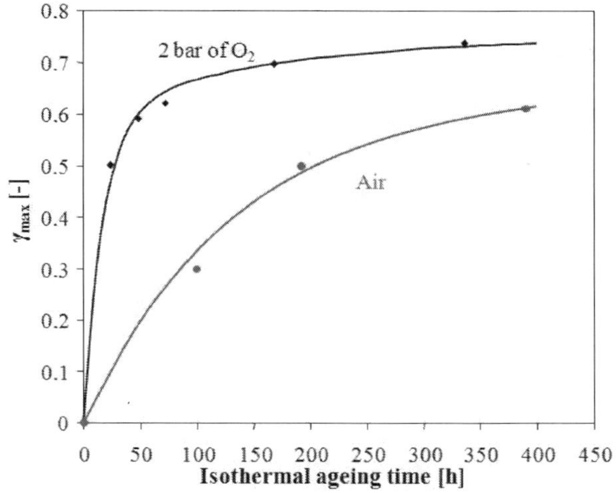

Figure 1.19. *Value of γ (γ_{max}), measured at 40 µm from the external edge as a function of the oxidation duration, for the TACTIX resin aged at 150°C in air at atmospheric pressure or under 2 bar O_2 (see also [GIG 16a, MIN 13, MIN 14]). For a color version of this figure, see www.iste.co.uk/gigliotti/ mechanical.zip*

Whatever the partial pressure of oxygen, the experimental values of γ_{max} can be mathematically expressed using a single expression:

$$\gamma_{max}(150°C, p, t) = 0.75 - 0.6\exp^{-\frac{t}{\tau_1}} - 0.15\exp^{-\frac{t}{\tau_2}} \qquad [1.4]$$

where τ_1 and τ_2 are two characteristic times dependent on the partial pressure of oxygen.

	Air	2 bar O_2	Air/2 bar O_2
τ_1 [h]	130	17.5	7.33
τ_2 [h]	1,200	164	7.32

Table 1.1. *Characteristic times τ_1 and τ_2 according to the partial pressure of oxygen (see also [GIG 16a, MIN 13, MIN 14])*

Table 1.1 shows that the relations $\tau_{1(air)}/\tau_{1(2\ bar\ O2)}$ and $\tau_{2(air)}/\tau_{2(2\ bar\ O2)}$ are very similar at approximately 7.3. Consequently, Figure 1.20 shows that if we plot the experimental values of γ_{max} according to the reduced ageing time $t^* = t \times \lambda(p)$, with $\lambda(p)$ equal to 1 for ageing in air and 7.3 for ageing under 2 bar O_2, they are situated along a "master curve" expressed analytically by:

$$\gamma_{max}(150°C, p, t) = 0.75 - 0.6\exp^{-\frac{t^*}{130}} - 0.15\exp^{-\frac{t^*}{1200}} \quad [1.5]$$

which proves the relevance of a time–pressure equivalence.

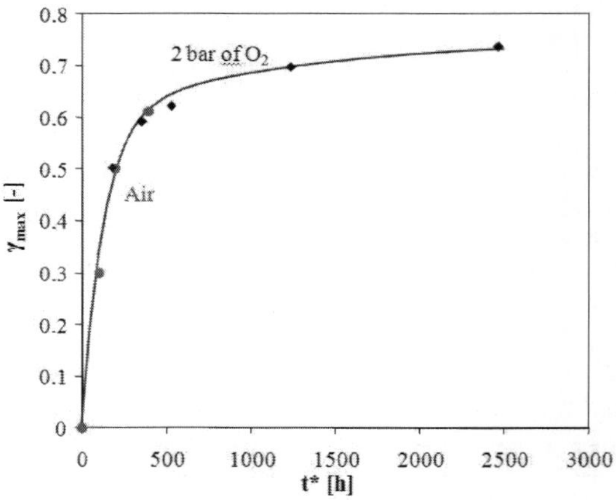

Figure 1.20. *Ageing "master curve", visualizing the evolution of γmax as a function of the reduced time t* (T = 150°C, $p_{ref.} = p_{atm.}$) (see also [GIG 16a, MIN 13, MIN 14]). For a color version of this figure, see www.iste.co.uk/gigliotti/mechanical.zip*

For example, from this curve, the γ_{max} value reached after ageing at 150°C for 2,500 hours in air at atmospheric pressure is equal to that measured after 360 hours under 2 bar O_2.

Figures 1.21 and 1.22 provide a link between the measurements of γ (see Figures 1.21(a) and 1.22(a)) and the indentation curves "load versus indentation depth" (see Figures 1.21(b) and 1.22(b)) for different ageing times under 2 bar O_2 at 150°C (see Figure 1.21) or, for different types of conditioning, at 150°C (in air or under 2 bar O_2, see Figure 1.22).

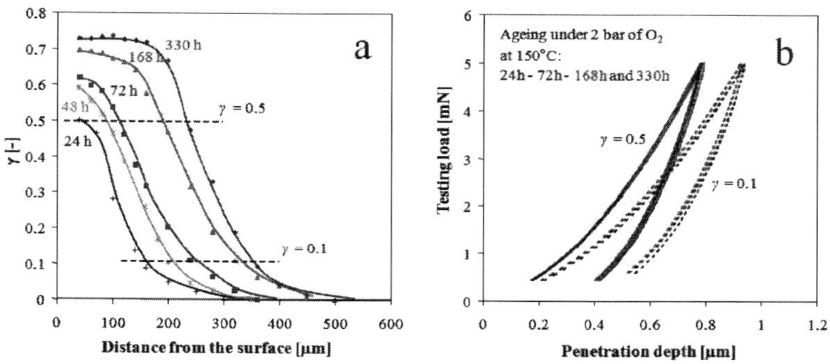

Figure 1.21. *a) Profiles of γ as a function of the distance from the edge and the oxidation time for the TACTIX resin aged under 2 bar of O_2 at 150°C. b) Comparison of indentation curves corresponding to two values of γ: 0.1 and 0.5 (see also [GIG 16a, MIN 13, MIN 14]). For a color version of this figure, see www.iste.co.uk/gigliotti/mechanical.zip*

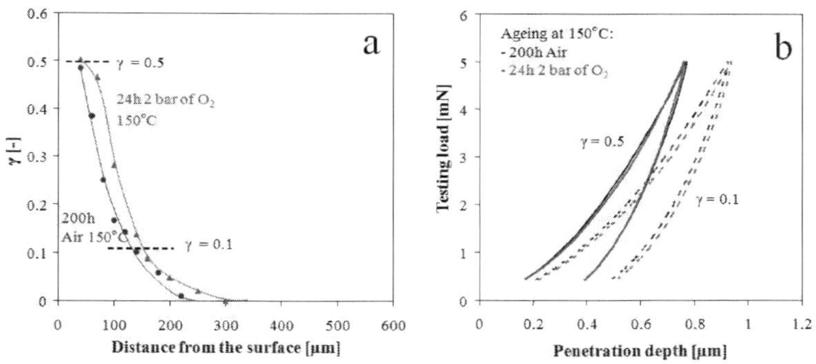

Figure 1.22. *a) Profiles of γ as a function of the distance from the edge of the TACTIX resin samples aged at 150°C, 200 hours in air at atmospheric pressure and 24 hours under 2 bar O_2. b) Comparison of indentation curves corresponding to two values of γ: 0.1 and 0.5 (see also [GIG 16a, MIN 13, MIN 14]). For a color version of this figure, see www.iste.co.uk/gigliotti/mechanical.zip*

In the case of ageing at 150°C in air at atmospheric pressure, Figure 1.21 shows that, whatever the ageing time, we obtain the same value of γ in the samples, and identical indentation curves are recorded. In Figure 1.21(b), we obtain, for both values of γ considered ($\gamma = 0.1$ and $\gamma = 0.5$), four identical indentation curves at four very different distances from the edge according to the oxidation time of the samples. The same can be observed when comparing the indentation curves obtained in the oxidized layers of aged samples in different environments (Figure 1.22): the same value of γ corresponds to the same indentation curve. These observations suggest that there may be a direct link between the local mechanical behavior of the polymer and the phenomenological ageing parameter.

In conclusion, the phenomenological parameter γ – measured experimentally from the unloading phase of the indentation curves and using the method of Oliver and Pharr – can be considered both as a "tracer" of oxidation chemistry, given the existence of a correlation between γ and the concentration of the oxidation products, and as an indicator of the local mechanical behavior of the material. However, the parameter γ, estimated in this way, is a parameter that only accounts for the "elastic" part of the polymer behavior, whereas the latter is more complex and, in particular, time dependent.

1.1.4. *Characterizing the local mechanical behavior of the polymer*

In order to characterize the local behavior of the polymer more finely, a new experimental approach has been developed (see Figure 1.23). This approach is based on the joint use and analysis of measurements by ultra-micro-instrumented indentation (UMI) and by interferometric (IM) microscopy. This includes:

– recording the indentation curve "load versus indentation depth", during an UMI test (see Figure 1.23(a)). This curve allows the residual indentation print to be measured after unloading: the dimensions of this print depend on the parameters of the UMI test and on the local ageing state of the polymer;

– monitoring the shape and dimensions of the indentation prints using IM measurements carried out at regular intervals between 10 minutes and 3 months after the indentation test. The IM provides a 3D reconstruction of the print (see Figure 1.23(b)), which can also be exploited by following the evolution over time either of a profile (see Figure 1.23(c)) or of a particular point of the indented surface (see Figure 1.23(d)).

This experimental technique is highly innovative: the two tests provide additional information on the local mechanical behavior of the polymer, in the sense that the UMI test reveals the behavior of the material in the short term, while the IM adds information about the behavior of the polymer in the long term. The analysis of the experimental results allows the ingredients necessary for the construction of a constitutive law of virgin and aged polymers to be specified. The stages of setup, development and identification of this constitutive law will be detailed in the second and third chapters of this book.

Figure 1.24 shows 15 indentation curves obtained for the virgin material with a maximum load of 5 mN and a loading rate of 0.25 mN/s, without a holding phase: the 15 curves are superimposed, revealing that the initial mechanical properties of the resin in the virgin state are homogeneous. The scatter is very small: the indentation depth at the maximum load has a standard deviation of approximately 0.015 μm (for an average value of approximately 1 μm), showing that the UMI tests are very reproducible.

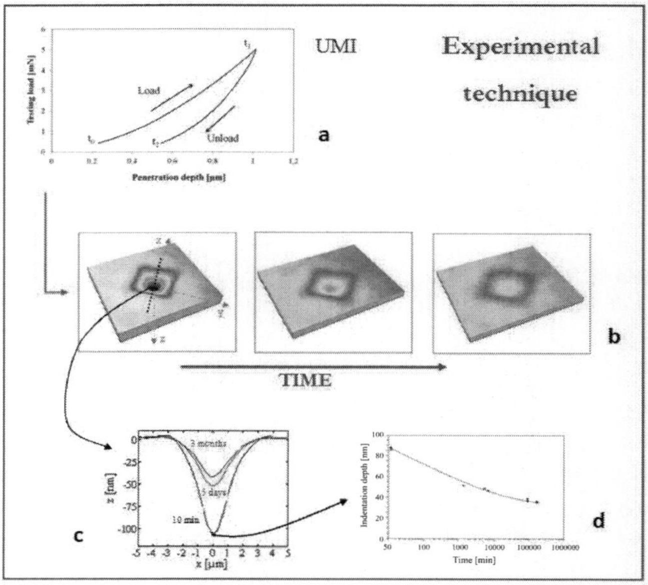

Figure 1.23. *Experimental technique based on the joint use of UMI to measure "load versus indentation depth" curves and IM to monitor the relaxation of the indentation print after unloading (see also [GIG 16a, MIN 13, MIN 14]). For a color version of this figure, see www.iste.co.uk/gigliotti/ mechanical.zip*

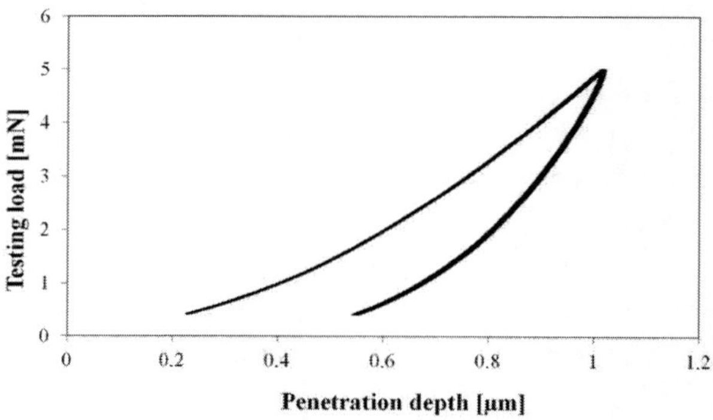

Figure 1.24. *"Load versus indentation depth" curve obtained by UMI on virgin TACTIX resin: 15 indentation curves are superimposed (see also [GIG 16a, MIN 13, MIN 14])*

By always imposing the same maximum load of 5 mN, Figures 1.25 and 1.26, respectively, show the indentation curves obtained with different loading speeds (from 0.12 to 0.25 mN/s) and with different holding times of the maximum load (from 20 seconds to 60 seconds): in the speed range analyzed, the loading rate does not influence the indentation curve (see Figure 1.25), while the duration of the creep phase has a significant effect on the indentation depth, e.g. after 60 seconds of load holding, the indentation depth increases by 7%, from 0.991 μm to 1.058 μm (see Figure 1.26).

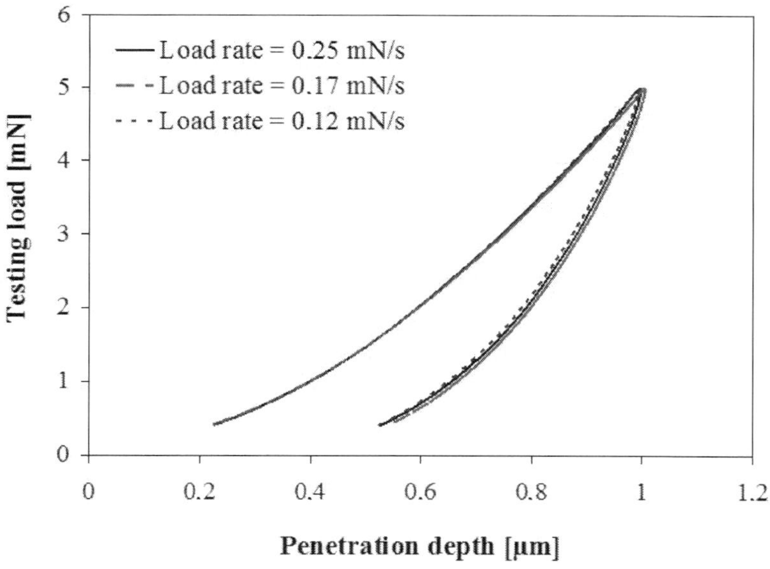

Figure 1.25. *"Load versus indentation depth" curve obtained by UMI on virgin TACTIX resin: effect of loading speed (see also [GIG 16a, MIN 13, MIN 14]). For a color version of this figure, see www.iste.co.uk/gigliotti/mechanical.zip*

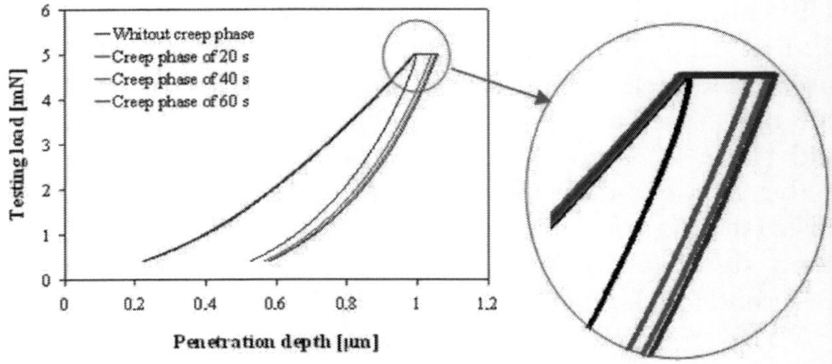

Figure 1.26. *"Load versus indentation depth" curve obtained by UMI on virgin TACTIX resin: effect of creep phase (see also [GIG 16a, MIN 13, MIN 14]). For a color version of this figure, see www.iste.co.uk/gigliotti/mechanical.zip*

In order to better characterize the behavior of the polymer, a "progressive" indentation test was set up. During this test, the indentation load varies over time to impose progressive loading/unloading with maximum loads ranging from 1 mN to 5 mN (see Figure 1.27).

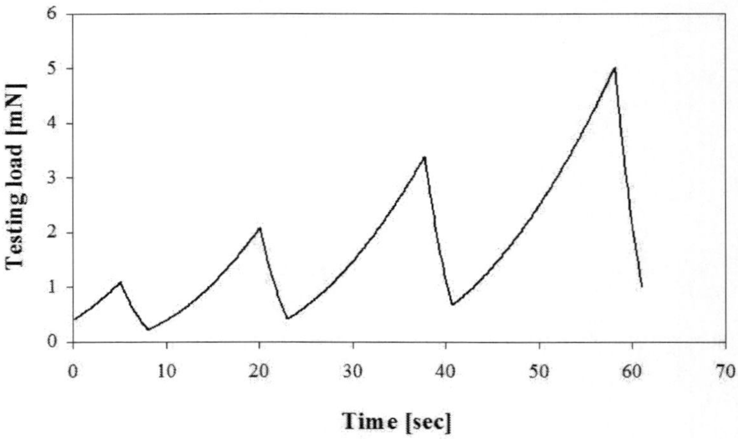

Figure 1.27. *Load imposed during a "progressive" indentation test (see also [GIG 16a, MIN 13, MIN 14])*

Figure 1.28 compares the "progressive" indentation curve with that obtained with a continuously increasing load ("monotonous" test). The two loading paths lead to a similar overall behavior and indentation depth values at maximum load. The presence of the loops – corresponding to the loading/unloading cycles – is representative of a viscous behavior of the polymer in the short term.

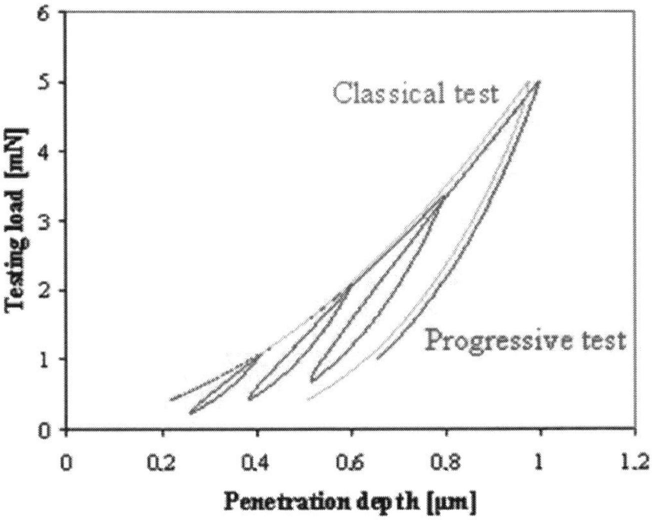

Figure 1.28. *Comparison of a "progressive" indentation test (red curve) with that of a "monotonous" test (grey curve) (see also [GIG 16a, MIN 13, MIN 14]). For a color version of this figure, see www.iste.co.uk/gigliotti/mechanical.zip*

The EIT modulus values, measured during unloading of the "progressive" indentation test as a function of the maximum values of the loading cycles, are plotted in Figure 1.29. The EIT values were found to decrease as the load increases, which suggests a "softening" behavior of the material during the test. The EIT values measured vary from approximately 3,500 MPa at the first loop (maximum load 1 mN) to approximately 3,200 MPa at the last loop (maximum load 5 mN). This final value approximates the EIT value obtained during the "monotone" indentation test.

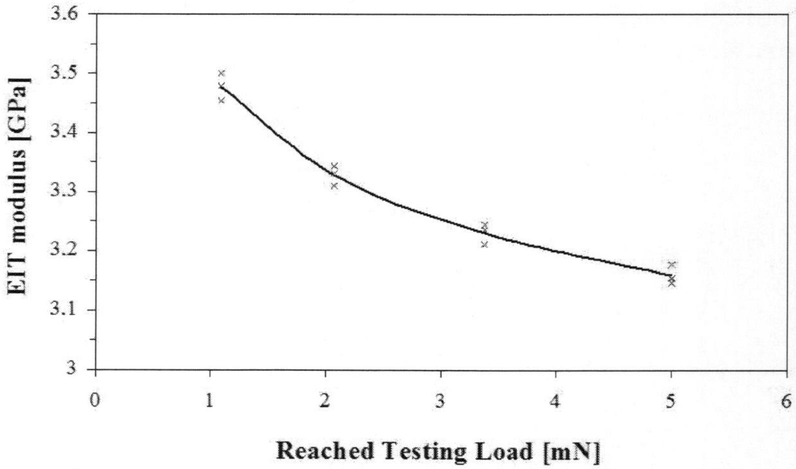

Figure 1.29. *Evolution of the EIT modulus during a "progressive" indentation test. Virgin TACTIX sample (see also [GIG 16a, MIN 13, MIN 14]). For a color version of this figure, see www.iste.co.uk/gigliotti/mechanical.zip*

Figure 1.30 shows the evolution of the morphology of an indentation print on the surface of the polymer. Figure 1.30(a), (b) and (c), respectively, illustrates the shape of the 3D print at approximately 10 minutes, 5 days and 3 months after the indentation test. From the 3D images, the indentation profiles along appropriate directions can be measured (Figure 1.30(d)) and their evolution can be characterized as a function of time. Figure 1.30(d) thus shows that the shape of the print is reasonably smooth and that the maximum indentation depth decreases with time from approximately 85 nm (10 minutes after the indentation test) to approximately 42 nm (after 3 months).

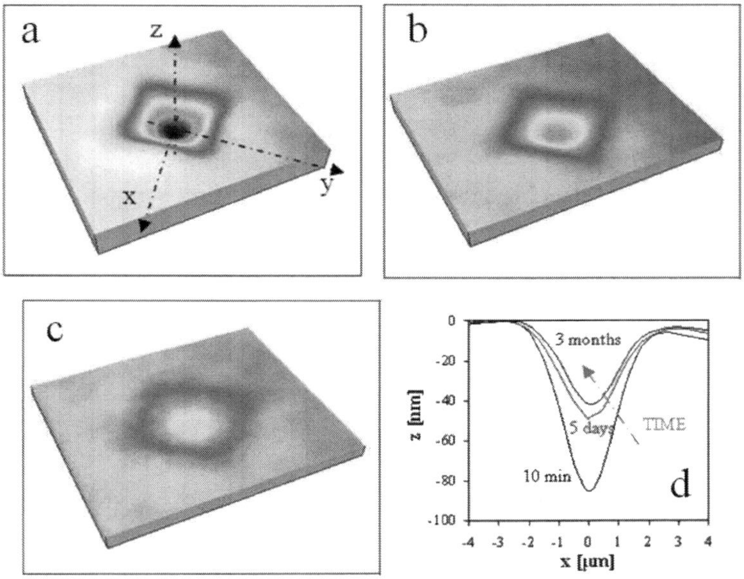

Figure 1.30. *Images by IM of an indentation curve on the surface of a virgin sample: a) 10 minutes; b) 5 days; c) 3 months after indentation test and d) evolution of a profile over time (see also [GIG 16a, MIN 13, MIN 14]). For a color version of this figure, see www.iste.co.uk/gigliotti/mechanical.zip*

The evolution of the maximum indentation depth as a function of time is illustrated in Figure 1.31: there is a continuous, significant decrease in recovery in the long term. The maximum depth decreases by more than 50% in 3 months: given the tendency of the recovery curve, it is viable to question whether or not a completely relaxed state is reached.

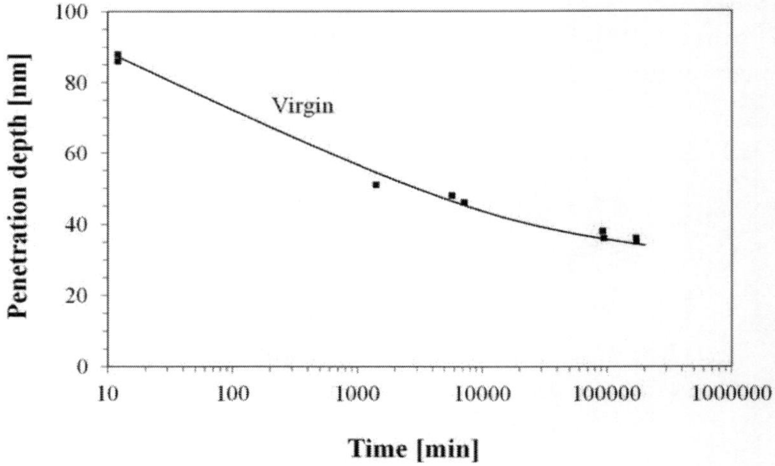

Figure 1.31. *Recovery of indentation curve after a UMI test: virgin TACTIX sample (see also [GIG 16a, MIN 13, MIN 14])*

Assuming that the recovery process is activated and accelerated by a suitable coupling of temperature and duration, a sample was placed in an oven at 150°C for 10 minutes under vacuum. This temperature (100°C below the glass transition temperature of the material) and the short duration of this conditioning were chosen to prevent rearrangement of the molecular structure, which occurs close to T_g, and the complementary physical ageing mechanisms; in addition, the experiment was carried out under vacuum to inhibit thermo-oxidation. Then, the surface of the sample was re-observed by IM at ambient temperature: the indentations disappeared, whereas the polishing scratches and surface defects are always visible. The same behavior was observed for temperatures below 150°C, but with longer holding times required to obtain complete recovery of the prints. The indentation test did not promote the occurrence of permanent strains in the material – at least for the retained maximum load level.

The main observations of this experimental study on virgin material are summarized below:

– viscous behavior is noted both in the short term – where two significant indices are the formation of loops during the progressive indentation test (see Figure 1.28) and the evolution of the indentation depth during the holding phase at maximum load (creep, see Figure 1.26) – and in the long term – where recovery is still active after 3 months (see Figure 1.31);

– a softening behavior of the polymer is suggested by the progressive indentation test (see Figure 1.29);

– the relaxation process leads to the complete recovery of the indentation prints, indicating the absence of permanent strains.

1.1.5. *Oxidized material*

Figure 1.32 shows the indentation curves obtained along the thickness of a sample aged for 120 hours at 150°C under 2 bar O_2. The maximum indentation depth – that is to say, the indentation depth at the maximum load of the test (5 mN) – increases when moving away from the surface exposed to the environment – that is to say, when a less oxidized material is considered. Thus, for example, a maximum indentation depth of 0.69 μm is measured at 20 μm from the edge, where there is maximum oxidation, whereas a value of 0.98 μm at 500 μm is measured inside the oxidized layer.

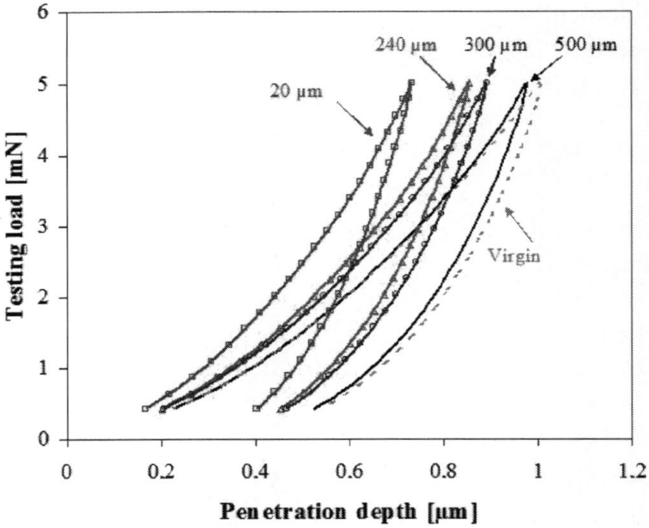

Figure 1.32. *Indentation curves at different distances from the edge of the oxidized sample for 120 hours at 150°C under 2 bar O_2 (see also [GIG 16a, MIN 13, MIN 14]). For a color version of this figure, see www.iste.co.uk/gigliotti/mechanical.zip*

Note that, in this figure:

– the indentation curve of the polymer at 20 µm from the surface exposed to the environment (green line with squares) is very different from that obtained at the center of the test piece (black line);

– the core of the sample (black line) exhibits a behavior similar to that observed for the virgin polymer (dotted line), which is consistent with the reaction/diffusion phenomenon characterizing thermo-oxidation.

Similar observations are recorded for the indentation curves carried out at a constant distance from the surface exposed to the environment for different ageing durations. For example, Figure 1.33 illustrates indentation measurements carried out at a constant distance of 20 μm from the surface directly exposed to the environment in samples aged for 24 hours, 72 hours and 120 hours: the maximal indentation depth decreases when the ageing time increases, from 0.99 μm (in the virgin state, continuous line) to 0.69 μm (after 120 hours of ageing, green line with squares).

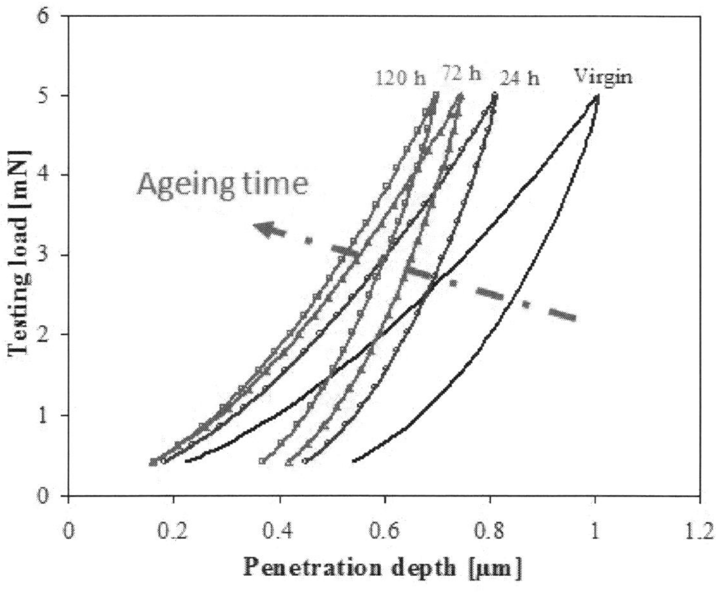

Figure 1.33. *Indentation curves at 20 μm from the edge of the sample. Black curve: virgin sample. Curves with symbols: samples oxidized at 150°C under 2 bar O_2 for 24 hours (circles), 72 hours (triangles) and 120 hours (squares) (see also [GIG 16a, MIN 13, MIN 14]). For a color version of this figure, see www.iste.co.uk/gigliotti/mechanical.zip*

The observations in Figures 1.32 and 1.33 suggest a change in the mechanical behavior of the polymer associated with its oxidation level. Qualitatively, the indentation curves of the oxidized material are less "open" and have a "steeper" unloading compared with the curves of the virgin material.

As for the virgin resin, the shape of the indentations on the surface of the samples was analyzed by IM at regular time intervals, from a few minutes after the UMI test to a maximum of three months.

The profile in Figure 1.30(d), measured 10 minutes after the indentation test on the surface of the virgin resin, is compared, in Figure 1.34, with a profile measured on a sample aged for 72 hours at 150°C under 2 bar O_2: the maximum indentation depth of approximately 51 nm is 40% less than the "virgin" value. However, the two profiles have a similar shape, with an identical internal slope (α in Figure 1.34), equal to 45 nm/μm.

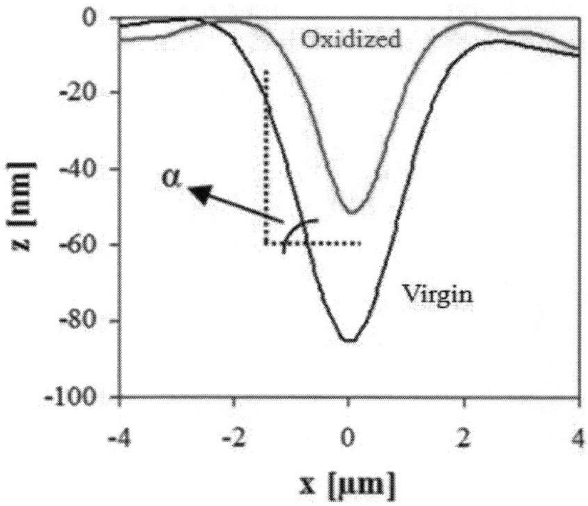

Figure 1.34. *Comparison of profiles of indentation curves measured on a virgin sample (black curve) and at 20 μm from the edge of a sample oxidized for 72 hours at 150°C under 2 bar O_2 (red curve) (see also [GIG 16a, MIN 13, MIN 14]). For a color version of this figure, see www.iste.co.uk/gigliotti/mechanical.zip*

The evolution of maximum indentation depth as a function of time is detailed in Figure 1.35; the circles represent the virgin state and the triangles the oxidized state (72 hours under 2 bar O_2, at 150°C). In Figure 1.35, there is a continuous decrease in the indentation depth over time, indicating viscous behavior in the long term, for both the virgin and oxidized polymers. The recovery curve of the oxidized resin is similar to that of the virgin resin, although clearly distinct.

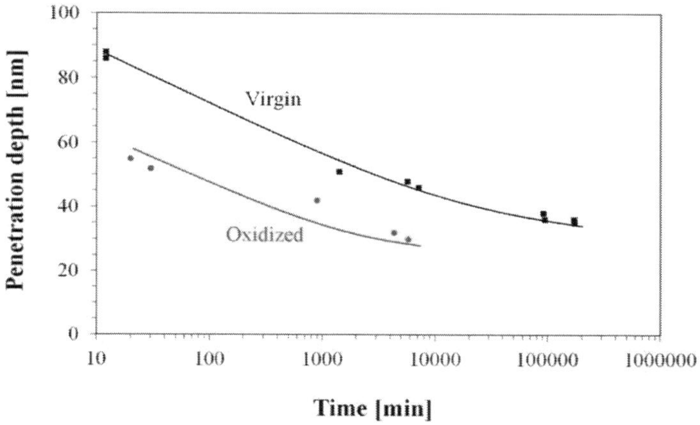

Figure 1.35. *Comparison between the recovery curve of a virgin sample (in black) and that obtained from a sample oxidized for 72 hours under 2 bar O_2 at 150°C (in red) (see also [GIG 16a, MIN 13, MIN 14]). For a color version of this figure, see www.iste.co.uk/gigliotti/mechanical.zip*

Owing to the complexity of the experimental results, implementing and developing a numerical model is essential to give a correct interpretation of the local mechanical behavior of virgin and oxidized polymers. This aspect will be developed in the second and third chapters of this book.

1.2. Study of matrix shrinkage induced by thermo-oxidation in unidirectional OMCs

The reference [VU 12] presents a number of images obtained by scanning electron microscopy (SEM), which shows the presence of matrix shrinkage and debonding at the fiber/matrix interfaces in UD composites (see Figure 1.36).

The experimental study was carried out on a UD sample $[0]_{50}$ (50 plies) made of carbon composite/epoxy HTS/TACTIX with the dimensions 15 × 15 × 10 mm, according to the protocol described in the introduction. The position on the surface observed is marked, with respect to a corner of the surface, with the IM apparatus being able to automatically move along the x and y axes in a very reproducible manner. In this case, along each axis, a movement of 2 mm has been retained, in order to reasonably disregard the edge effects.

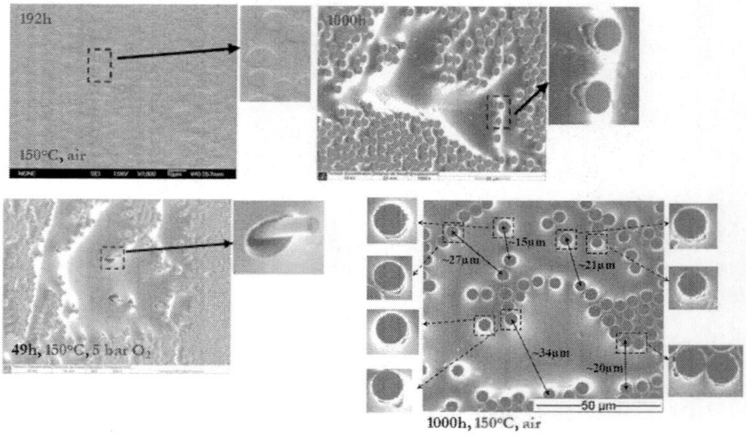

Figure 1.36. *Observations using SEM, which reveal the presence of matrix shrinking and debonding at the fiber/matrix interfaces in oxidized UD composites [VU 12]*

To more easily distinguish the matrix fibers, only a part of the image obtained by IM of dimensions 100 × 100 μm will be shown, as shown in Figure 1.37. However, all measurements will be relative to the entire original surface, with dimensions of 350 × 350 μm, to have the widest possible range of experimental data for the implementation of a statistical study.

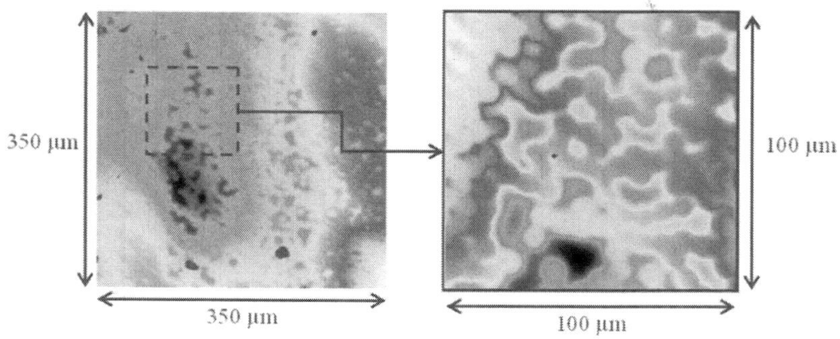

Figure 1.37. *Part of an IM image showing the development of matrix shrinkage associated with the occurrence of thermo-oxidation phenomena (see also [GIG 16b, GIG 16c]). For a color version of this figure, see www.iste.co.uk/gigliotti/mechanical.zip*

In these images, the carbon fibers are easily distinguishable from the matrix. The surface observed is orthogonal to their direction, and the heads of the fibers appear circular with an average diameter of 7 μm, a characteristic value for this type of fiber. Between two fibers, the matrix shrinks, as shown in Figure 1.38. The distance between fibers is defined as the distance between their axes of symmetry along the x axis and the depth of the shrinkage is measured along the z axis. The shrinkage depth varies along the x axis to its maximum value located close to the center. In the same figure, the z axis allows the distance to the surface directly exposed to the environment to be identified, according to which all gradients associated with thermo-oxidation (concentration of O_2, concentration of oxidation products, properties, etc.) develop.

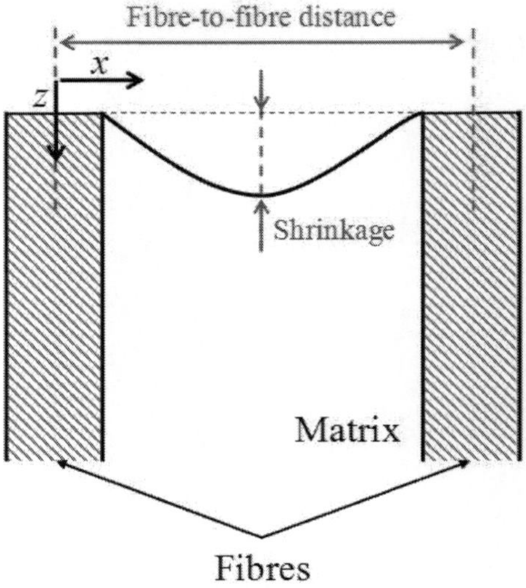

Figure 1.38. *Schematic of matrix shrinkage between two fibers, associated with thermo-oxidation (see also [GIG 16b, GIG 16c]). For a color version of this figure, see www.iste.co.uk/gigliotti/mechanical.zip*

1.2.1. *Virgin sample*

The virgin UD composite has, on the surface, slight matrix shrinkage, associated, on the one hand, with the presence of residual strains of thermal origin, due to the cooling of the sample from curing temperature to ambient temperature, and, on the other hand, to the development of permanent strains due to the polishing phase. With regard to the latter, a preliminary study was carried out to check the influence of polishing on matrix shrinkage measurements. In particular, three parameters were taken into account: the force on the sample holder, the duration of each polishing step and the lubricants. The influence of the supporting force and the polishing time on matrix shrinkage is negligible, whereas the use of alcohol-based lubricants leads to superficial damage and to the progressive shrinkage increase. We have therefore chosen to apply the same protocol implemented for the polishing of the pure resin, which ensures a clean composite surface and does not affect the integrity of the matrix or the intensity of the matrix shrinkage. The IM measurement of the surface of a virgin sample (see Figure 1.39) shows that:

– the arrangement of the fibers in the composite is not uniform since matrix-rich zones are observed. For example, the four identified areas (A, B, C and D) have local fiber volume fraction (V_f) values that vary between 0.64 – a value close to the nominal fiber volume fraction of the plate – and 0.34 (all values are shown in Table 1.2);

– in all these areas, a pair of fibers at a distance of approximately 20 can be found;

– around each pair, the neighboring fibers can be distributed randomly, having tight or dispersed configurations.

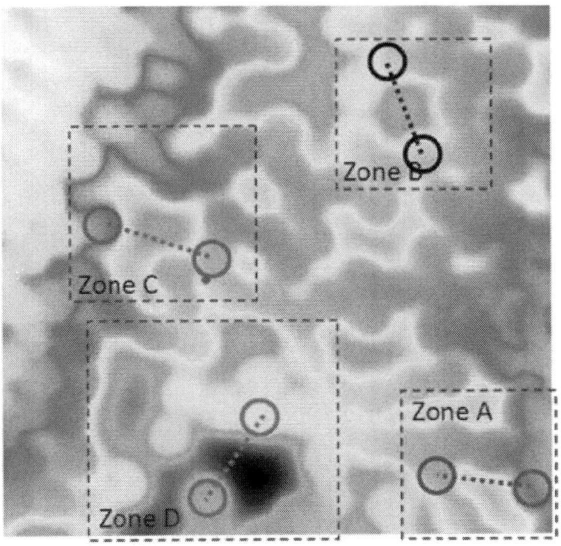

Figure 1.39. *IM image of the surface of a virgin UD sample (see also [GIG 16b, GIG 16c]). For a color version of this figure, see www.iste.co.uk/gigliotti/mechanical.zip*

	Zone A	Zone B	Zone C	Zone D	Nominal value of the plate
V_f [-]	0.64	0.50	0.34	0.36	0.66

Table 1.2. *Local fiber volume fraction in zones A, B, C and D and the nominal value of the plate (see also [GIG 16b, GIG 16c])*

Figure 1.40 shows that the matrix shrinkage profile associated with each pair of fibers identified in Figure 1.39 is different: area A (fiber-rich zone) has a smaller matrix shrinkage than zone D (matrix-rich zone).

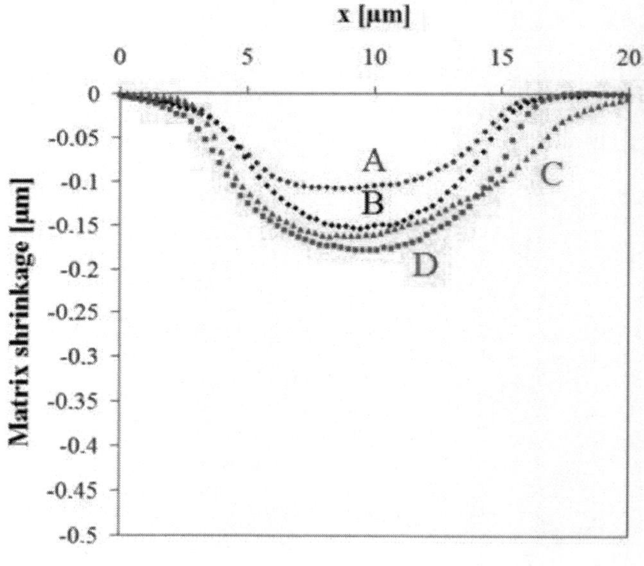

Figure 1.40. *Matrix shrinkage for a distance of 20 µm between fibers. Virgin UD sample (see also [GIG 16b, GIG 16c]). For a color version of this figure, see www.iste.co.uk/gigliotti/mechanical.zip*

Moreover, the maximum shrinkage can be correlated with the associated V_f (see Figure 1.41), demonstrating that, for a given distance between fibers, the matrix shrinkage is a function of the local distribution of the fibers: for a V_f equal to 0.34, the maximum matrix shrinkage is approximately 80% greater than that measured for a V_f of approximately 0.64.

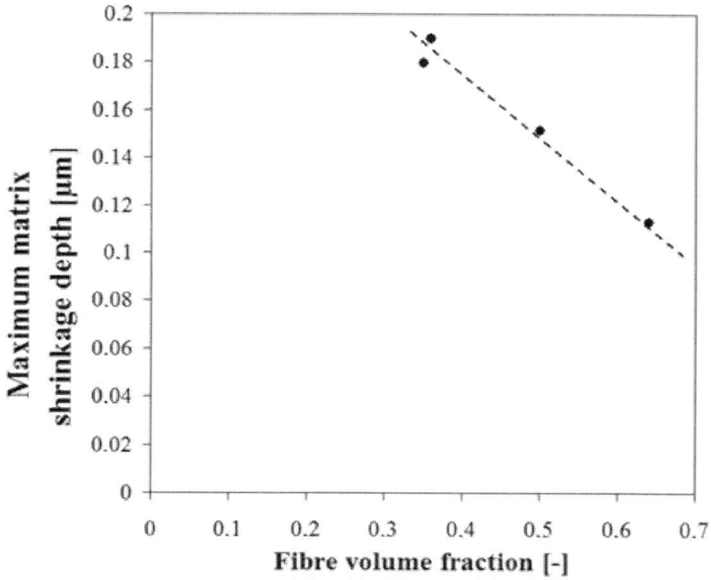

Figure 1.41. *Maximum matrix shrinkage as a function of the fiber volume fraction, for a distance of approximately 20 µm between fibers. Virgin UD sample (see also [GIG 16b, GIG 16c])*

In Figure 1.42, the values of the maximum matrix shrinkage were plotted as a function of the distance between fibers: the matrix shrinkage increases with the distance between fibers and the measurements are significantly dispersed. To understand the origin of this scatter, in Figure 1.42, the maximum matrix shrinkage values associated with the four profiles A, B, C and D were localized. The four red dots fit well into the experimental cloud and the difference between the extreme values defines the width of the scatter band. This allows us to associate the scatter of the measurements with that of the local configurations of fibers.

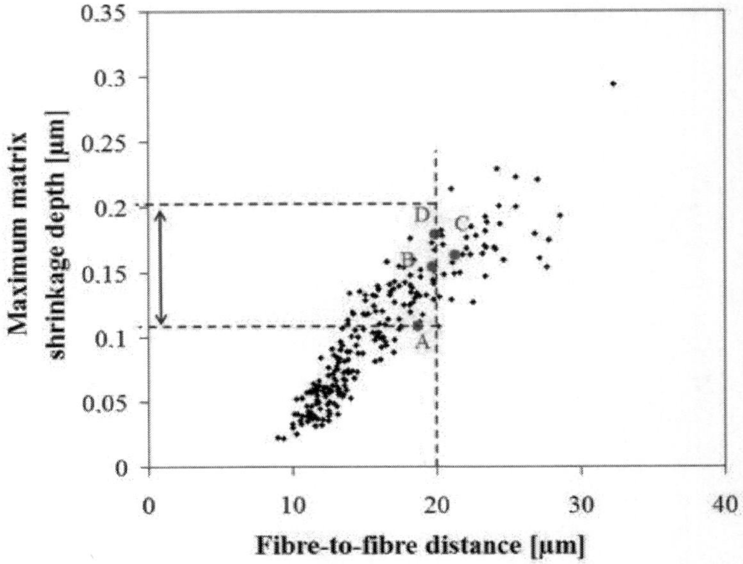

Figure 1.42. *Maximum matrix shrinkage as a function of the distance between fibers. Virgin UD sample (see also [GIG 16b, GIG 16c]). For a color version of this figure, see www.iste.co.uk/gigliotti/mechanical.zip*

The experimental study on the virgin composite thus concludes that local matrix shrinkage changes are related to a "structural" effect: with the fibers being more rigid than the matrix, their distribution in the composite affects the local mechanical strains, leading to different shrinkage values depending on the distance between fibers and the local volume fraction (V_f). These two parameters are related to the random distribution of fibers in the composite, which can only be studied statistically. For example, Figure 1.43 shows the probability that two fibers are at a given distance.

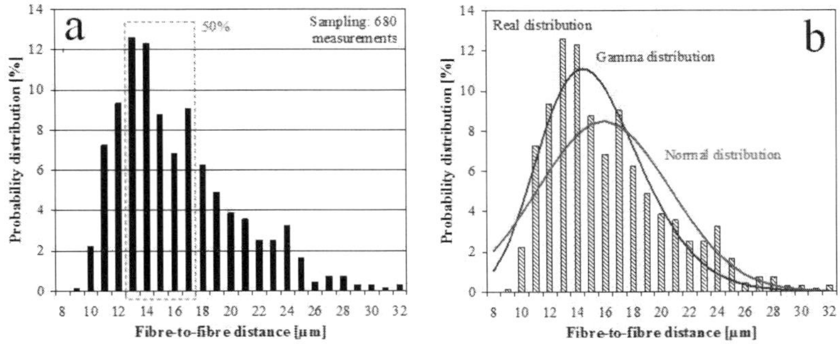

Figure 1.43. *a) Probability of finding two fibers at a distance between 8 µm and 32 µm on the surface of a UD sample. b) Comparison between a normal distribution law (green curve) and a Gamma law (blue curve) (see also [GIG 16b, GIG 16c]). For a color version of this figure, see www.iste.co.uk/gigliotti/mechanical.zip*

The following can be observed:

– 50% of the distances between fibers (approximately 680) are between 13 µm and 17 µm (see Figure 1.43(a));

– a normal distribution, with $\mu = 16$ (mean) and $\sigma = 4.7$ (standard deviation), does not correctly describe the actual distribution (see Figure 1.43(b));

– a gamma distribution, with $k = 17.3$ and $\theta = 1.13$, nicely follows a real distribution (see Figure 1.43(b)).

The same distribution can also be expressed in terms of cumulative probability (see Figure 1.44). The term "cumulative" indicates the probability that two fibers are at a distance that is smaller than a predetermined value, e.g. for the studied material, the fibers are at a distance of less than 12 μm in 20% of cases, less than 15 μm in 50% of cases and less than 19 μm in 80% of cases. In addition, the "complementary cumulative" value, which accounts for the probability of two fibers being at a greater distance than the predetermined distance, can be defined.

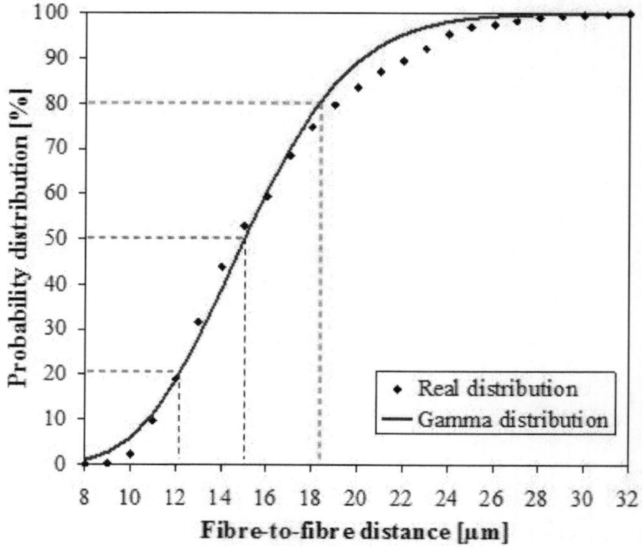

Figure 1.44. *Cumulative probability of having a distance between fibers less than a predetermined value (see also [GIG 16b, GIG 16c]). For a color version of this figure, see www.iste.co.uk/gigliotti/mechanical.zip*

In Figure 1.45(a), the width of the scatter, defined by the extreme measurement values, is represented for the fiber distances of 12 μm, 15 μm and 19 μm, corresponding, respectively, to a cumulative probability of 20%, 50% and 80%. In Figure 1.45(b), for these same values, the "average" shrinkage, s*, will be used hereafter to characterize the matrix shrinkage field on the surface of the UD composite.

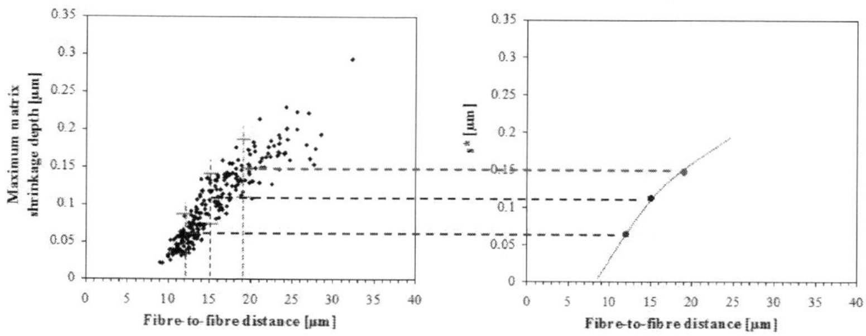

Figure 1.45. *a) "Maximum matrix shrinkage versus distance between fibers" curve for the virgin state. b) "Mean" shrinkage according to the distance between fibers (see also [GIG 16b, GIG 16c]). For a color version of this figure, see www.iste.co.uk/gigliotti/mechanical.zip*

1.2.2. *Sample oxidized under 2 bar O_2*

The sample, characterized previously as the initial virgin state, was oxidized in a climatic chamber at 150°C under 2 bar O_2. The initial measurements of the surface are carried out after 24 hours of ageing and then every two additional hours of ageing up to 42 hours because, at this point, the sample was already sufficiently damaged. Figure 1.46 shows the evolution of matrix shrinkage observed experimentally in the virgin state and for three ageing durations (24 hours, 38 hours and 42 hours). All images have the same scale (the vertical displacement is between 0 and 1 μm).

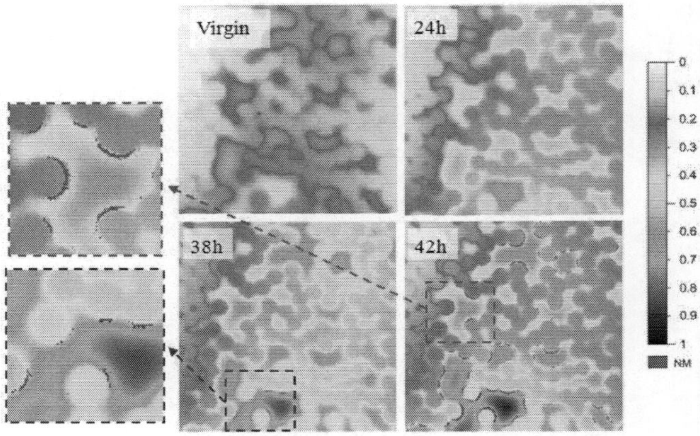

Figure 1.46. *Evolution of matrix shrinkage during isothermal ageing at 150°C under 2 bar O_2 (see also [GIG 16b, GIG 16c]). For a color version of this figure, see www.iste.co.uk/gigliotti/mechanical.zip*

In the initial state, the matrix shrinkage is small and the image is pink/red/yellow. After one day of ageing, light blue zones appear on the surface, indicating an increased matrix shrinkage. This evolution progresses during the oxidation, as revealed by the presence of dark blue zones after 42 hours of ageing. After 34 hours of ageing, unmeasured spots (visible in violet on the zoomed images) are located at the interfaces of the fibers in matrix-rich regions (where V_f is small). These points have already been associated by Vu *et al.* [VU 12] with the fiber/matrix debonding observed by SEM. The image obtained after 42 hours of ageing shows that the debonding sites, which initially only concern zones with small V_f, multiply and propagate on the surface, also affecting other areas with higher V_f. During thermo-oxidation, the change in matrix shrinkage of the four previously selected profiles A, B, C and D (distance between fibers of approximately 20 μm) is presented in Figure 1.47. The matrix shrinkage increases proportionally: after 30 hours of ageing, all the shrinkage is almost double, and after 42 hours, it is 250% higher than the corresponding initial values. After 42 hours of ageing, profiles C and D have zones with unmeasured points, unlike profiles A and B. This suggests, once again, that debonding is not correlated solely with the distance between fibers, but also with their local configuration.

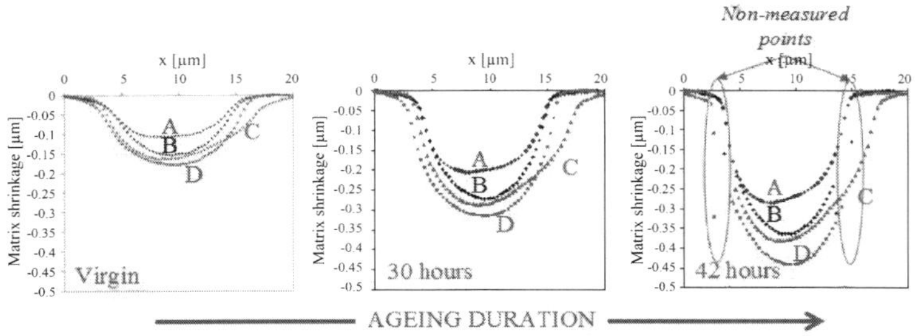

Figure 1.47. *Evolution of profiles A, B, C and D during thermo-oxidative ageing (at 150°C under 2 bar O_2) (see also [GIG 16b, GIG 16c]). For a color version of this figure, see www.iste.co.uk/gigliotti/mechanical.zip*

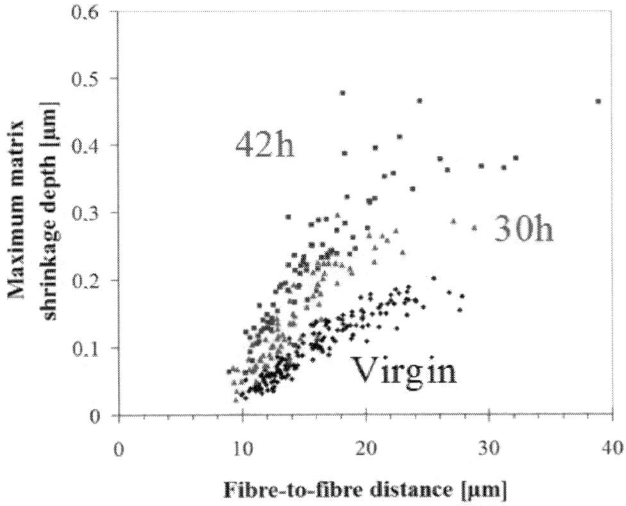

Figure 1.48. *Maximum matrix shrinkage according to the distance between fibers and oxidation time, at 150°C under 2 bar O_2 (see also [GIG 16b, GIG 16c]). For a color version of this figure, see www.iste.co.uk/gigliotti/mechanical.zip*

In Figure 1.48, the maximum matrix shrinkage is plotted as a function of the distance between fibers and the oxidation duration. The measurements show, on the one hand, that the matrix shrinkage increases with the distance between fibers and ageing time and, on the other hand, that the scatter of the measurements increases greatly with the level of oxidation. For example, the blue dots, representing the most oxidized state in this study, are the most dispersed, in particular for a distance between fibers of approximately 20 μm. In fact, when fiber/matrix debonding begins, new surfaces directly exposed to the environment are available, immediately becoming preferred paths for oxygen diffusion in the composite [COL 12], thereby accelerating ageing of the material. As a result, the increase in matrix shrinkage in these damaged areas will be faster compared with those where there is still no damage.

At the same time, the initiation of debonding promotes a variation in the local strains due to a reduction in the fiber constraint effect, also affecting the measured matrix shrinkage value. For a given oxidation time and distance between fibers, the local configuration may or may not lead to ruptures of the fiber/matrix interface. The sharp increase in shrinkage when debonding occurs explains the greater variability in the measurements after this damage occurs (see Figure 1.48). Experimental observations again lead to the conclusion that matrix shrinkage changes locally because of a "structural" effect, associated with the presence of fibers, and also depends on the oxidation level of the surface. Similarly to what was done in the previous section, the evolution of the matrix shrinkage field during oxidation can be represented by the corresponding parameter s* (see Figure 1.49).

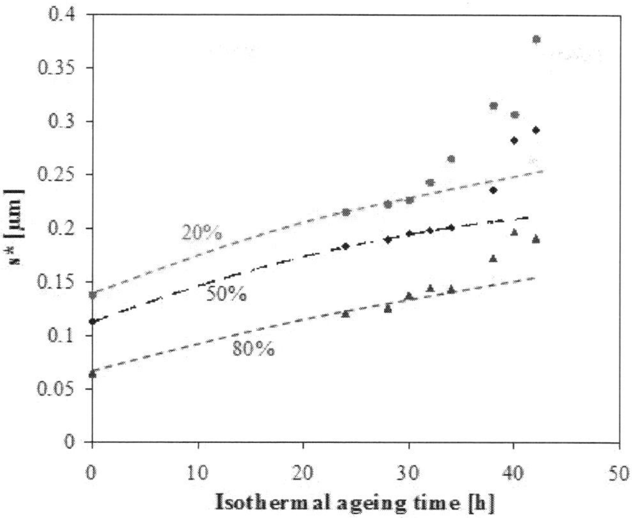

Figure 1.49. *Evolution of mean matrix shrinkage as a function of ageing time (150°C, 2 bar O_2) for three distances between fibers: 12 μm (blue), 15 μm (black) and 19 μm (red). The percentages indicate the probabilities of having two fibers at distances greater than the selected value (see also [GIG 16b, GIG 16c]). For a color version of this figure, see www.iste.co.uk/gigliotti/mechanical.zip*

Two regimes can be identified:

– a "pre-debonding" regime, where the evolution of s* follows a regular pattern (dashed curves in Figure 1.49);

– a "post-debonding" regime, combined with a more rapid increase in the mean matrix shrinkage.

In Figure 1.50, the oxidation time is replaced by the maximum value of the phenomenological ageing variable γ_{max}, since it is this value that characterizes the level of oxidation on the surface where debonding appears. With this representation, we clearly identify the two regimes, pre- and post-debonding. In the pre-debonding regime, the matrix shrinkage, s*, increases linearly with γ_{max}, up to a critical value (γ_{maxc}). Beyond this critical value, s* has a much faster evolution, which characterizes the post-debonding regime. The critical value of the phenomenological parameter γ_{maxc} does not vary much with the distance between fibers (cumulative probability). The value of γ_{maxc} of 0.56 represents a threshold between the healthy state and the damaged state.

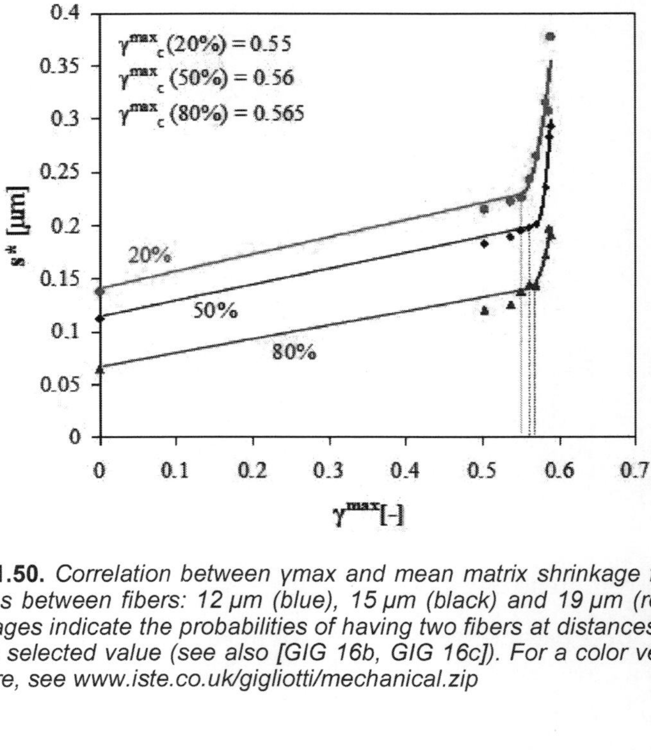

Figure 1.50. *Correlation between γmax and mean matrix shrinkage for three distances between fibers: 12 µm (blue), 15 µm (black) and 19 µm (red). The percentages indicate the probabilities of having two fibers at distances greater than the selected value (see also [GIG 16b, GIG 16c]). For a color version of this figure, see www.iste.co.uk/gigliotti/mechanical.zip*

In conclusion, ageing at 2 bar O_2 at 150°C shows that matrix shrinkage increases as a function of the degree of surface oxidation (γ_{max}) and that, starting from a critical value of γ_{max} (γ_{maxc} equal to 0.55 ± 0.01, i.e. 38 hours of ageing under 2 bar O_2), debonding at fiber/matrix interfaces appears on the surface of the sample, initially located in the matrix-rich parts. The statistical study makes it possible to visualize quite clearly the existence of two different regimes, pre- and post-debonding, and the critical value of γ_{max}.

1.2.3. *Air-oxidized sample*

A new ageing test was carried out on a UD composite sample in air at atmospheric pressure at 150°C for 14 days (instead of 2 days under 2 bar O_2). This will be called sample "B" and the one oxidized previously under oxygen pressure will be sample "A". Note: Both samples were initially polished rigorously with the same protocol described in the previous sections. In Figure 1.51, the maximum matrix shrinkage of the virgin sample B is plotted as a function of the distance between fibers and compared with the result obtained for the virgin sample A: in both cases, the experimental points have the same shape and the same scatter, proving that the initial state of the two samples is the same (same plate and same polishing protocol).

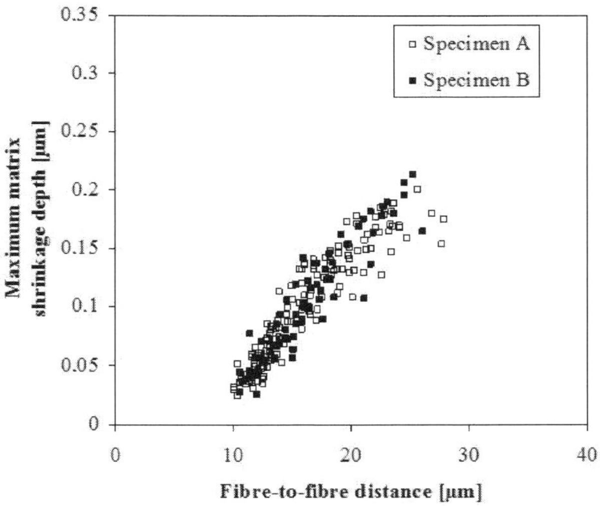

Figure 1.51. *Maximum matrix shrinkage as a function of the distance between fibers of virgin sample A (white squares) and virgin sample B (black squares) (see also [GIG 16b, GIG 16c])*

Figure 1.52 shows the evolution of the surface matrix shrinkage of sample B during oxidation. All images are on the same scale, between 0 μm (red) and 1.2 μm (dark blue).

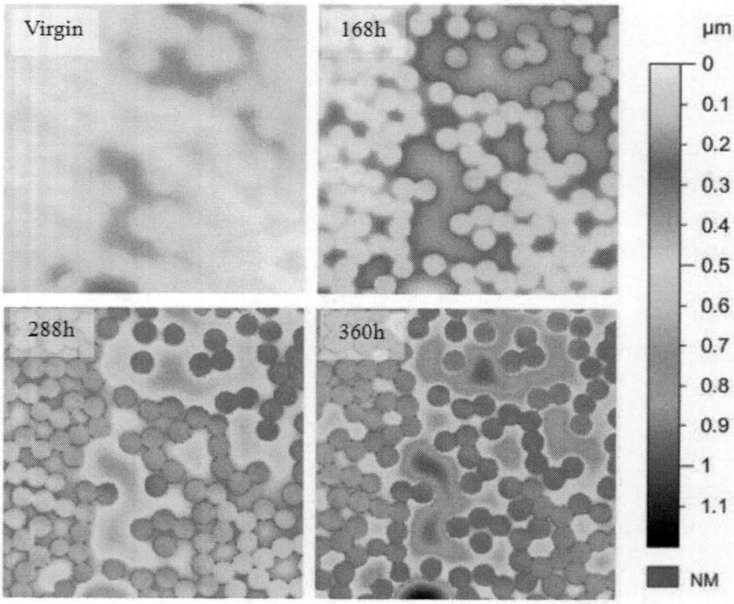

Figure 1.52. *Evolution of matrix shrinkage during isothermal ageing at 150°C in air at atmospheric pressure (see also [GIG 16b, GIG 16c]). For a color version of this figure, see www.iste.co.uk/gigliotti/mechanical.zip*

The degradation mechanisms are the same as under oxygen pressure:

– in the initial state, the matrix shrinkage is small and the image is pink/red;

– after 168 hours of ageing (7 days), yellow areas appear on the surface, indicating an increase in matrix shrinkage. This evolution progresses during oxidation. After 360 hours of ageing (15 days), there are dark blue colored areas;

– from approximately 288 hours of ageing, unmeasured points appear on the surface, which are localized at the interfaces of the fibers surrounding the matrix-rich regions (where the V_f are smaller), and where the distances between fibers are the greatest.

In Figure 1.53, the maximum matrix shrinkage as a function of the distance between fibers and the duration of oxidation is illustrated for sample A aged under 2 bar O_2 and for sample B aged in air at atmospheric pressure at the same temperature of 150°C. The matrix shrinkage of sample A after 24 hours of ageing (or 42 hours for Figure 1.53(b)) is perfectly superimposed on measurements taken on the surface of sample B after 168 hours of ageing (7 × 24) or 288 (7 × 42) hours for Figure 57(b). It would seem that the acceleration factor of the oxidation processes at 150°C, $\lambda \approx 7$ – found for the resin studied, between 2 bar O_2 and air at atmospheric pressure (see Table 1.1) – also applies to degradation processes of the corresponding UD composite.

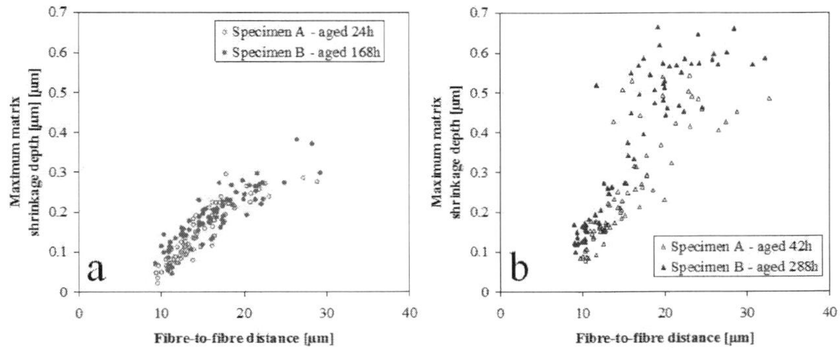

Figure 1.53. *Comparison of maximum matrix shrinkage as a function of the distance between fibers of sample A (unfilled symbols), oxidized at 150°C under 2 bar O_2 at a) 24 hours and b) 42 hours, and sample B (filled symbols), oxidized at 150°C in air at atmospheric pressure at a) 168 hours and b) 288 hours (see also [GIG 16b, GIG 16c]). For a color version of this figure, see www.iste.co.uk/gigliotti/mechanical.zip*

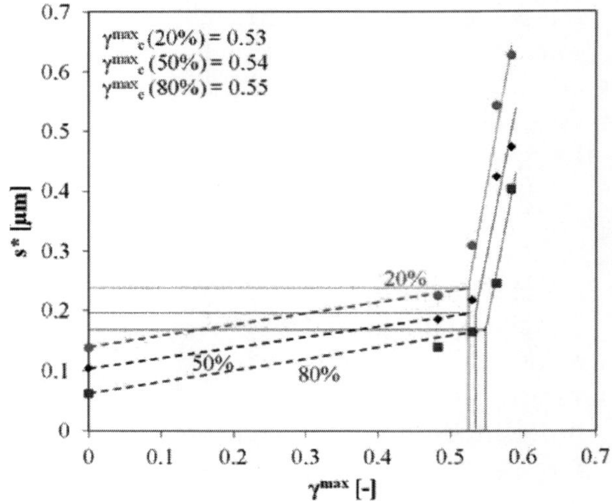

Figure 1.54. *Ageing at 150°C in air at atmospheric pressure: mean matrix shrinkage as a function of the distance between fibers and γ_{max}. The percentages indicate the probabilities of having two fibers at distances greater than the selected value (see also [GIG 16b, GIG 16c]). For a color version of this figure, see www.iste.co.uk/gigliotti/mechanical.zip*

In the same way as for sample A, the values of the average matrix shrinkage of sample B are plotted in Figure 1.54 according to the phenomenological parameter of surface ageing (γ_{max}), for the three distances: 12 µm, 15 µm and 19 µm – corresponding, respectively, to a cumulative probability of 20%, 50% and 80%.

Here again, the pre- and post-debonding regimes are clearly identifiable with critical values of γ_{max} very close to those obtained for ageing under oxygen pressure. Moreover, if the average shrinkage values are plotted, s* corresponding to 50% cumulative probability as a function of γ_{max} for the ageing at 150°C (A under 2 bar O_2 and B in air at atmospheric pressure), it appears that all points belong to the same curve (see Figure 1.55).

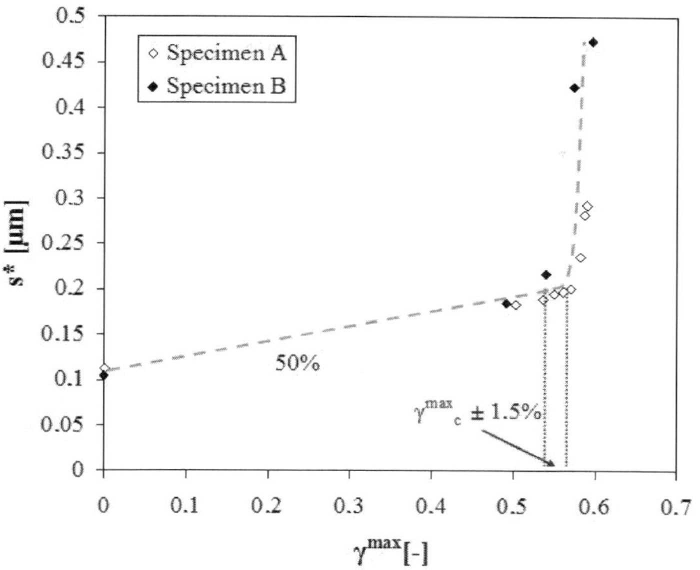

Figure 1.55. *Oxidation in air at atmospheric pressure and under oxygen pressure: mean matrix shrinkage as a function of γ_{max} for a 15 µm distance between fibers (50% cumulative probability) (see also [GIG 16b, GIG 16c]). For a color version of this figure, see www.iste.co.uk/gigliotti/mechanical.zip*

The critical value of γ_{max} appears in correspondence with the occurrence of debonding on 50% of the surface and leading to a sharp increase in the mean maximum shrinkage depths, within ± 1.5%, and is independent of the type of ageing considered. It can be therefore deduced that the evolution of matrix shrinkage at the surface and the onset of debonding depends solely on the surface ageing kinetics of the material. It has already been shown that, within the sample, the property gradients do not develop in the same way in atmospheric air as under 2 bar O_2, but, despite this, it is appropriate to have the same rate of surface oxidation (and hence the same γ value on the surface), so that the matrix shrinkage and the extent of the "damaged" surface are similar. The observations made in this section reinforce the effectiveness of a "moderate" pressure as an acceleration parameter of thermo-oxidation. In both environments, the matrix shrinkage shows a similar evolution with the surface oxidation rate, and in addition, debonding at the fiber/matrix interface occurs for a critical value, which appears to be characteristic of the material ($\gamma_{maxc} = 0.55 \pm 0.01$).

Again, the development and implementation of a numerical model is essential to give a correct interpretation of the complexity of the experimental results obtained for UD composite samples. This will be the objective of the second and third chapters of this book.

2

Modeling of Thermo-oxidative Ageing of OMCs

This chapter is devoted to the modeling of the local mechanical behavior of the polymer and the composite and includes:

– a short introduction to the thermodynamics of irreversible processes with internal variables (TIV);

– the development of an ageing-dependent behavior law for organic polymers;

– modeling and taking into account inelastic strains of chemical origin, associated with thermo-oxidation.

2.1. Thermodynamics of irreversible processes with internal variables

The modeling of the ageing of organic matrix composite materials is developed within the framework of thermodynamics of irreversible processes with internal variables (TIV) [ECK 40, DEG 62, BAT 79, GLA 71, MUL 73, MAU 99].

This modeling framework – a thorough and rigorous description of which can be found in work by De Groot and Mazur [DEG 62], and a critical approach in work by Truesdell [TRU 84] – postulates the existence of a transient state that tends towards a relaxed state, representing the equilibrium of the material. This "return to equilibrium" is achieved through a series of elementary mechanisms, described by internal variables, to which their own characteristic times are associated. Although the system is out of equilibrium, each elementary volume unit can be considered to be approximately in equilibrium from a thermodynamic point of view, thus thermodynamic and dissipation potentials can be defined at each time of the process.

The thermodynamic potential, the Helmholtz specific energy, ψ_R (free energy per unit mass), in a Lagrangian framework and with respect to an initial reference configuration, is a function of the state variables (V_e), internal variables (V_i) and, usually, temperature (T):

$$\psi_R = \psi_R\,(V_e, T, V_i) \qquad [2.1]$$

From the expression of this potential, the state laws can be described as follows:

$$F_e = \rho_R \frac{\partial \psi_R}{\partial V_e} \qquad [2.2]$$

$$F_i = -\rho_R \frac{\partial \psi_R}{\partial V_e}$$

$$s = -\frac{\partial \psi_R}{\partial T}$$

where ρ_R expresses the density in the reference configuration, F_e and F_i are the thermodynamic forces associated with, respectively, the state variables and the internal variables and s is the specific entropy. In the TIV approach, the thermodynamic potential, ψ_R, is often convex with respect to the state variables and the internal variables and concave with respect to temperature. It should also be noted that, in a Lagrangian formulation, the derivative of an objective operator is in turn an objective quantity.

During a dissipative process, the change in specific entropy of an elementary reference volume can be divided into two terms (see, e.g. [MUL 73]): one related to the entropy exchange with the external environment (to the thermodynamic system) and the other associated with the internal entropy production, due to dissipation (denoted $\sigma_V[s]$). According to the TIV approach, the latter term, sometimes referred to as intrinsic dissipation, is strictly positive in a dissipative process. The production of entropy and intrinsic dissipation can be related to purely thermal phenomena and, in this case, we refer to heat dissipation. The dissipation associated with the evolution of internal variables, denoted by Φ_{IV}, is usually written in TIV as the product of the generalized forces associated with the internal variables and the variation of such variables over time and can therefore be considered as a fictive power, as follows:

$$\rho_R \Phi_{IV} = T\sigma_V[s]_{IV} = \rho_R \sum_i F_i \frac{dV_i}{dt} \geq 0 \qquad [2.3]$$

This bilinear expression can be considered as a postulate under the TIV approach and as the core of all TIV models: it can be identically satisfied by defining a dissipation potential, $D(dV_i/dt)$, a function of the velocity of the internal variables, or its Legendre–Fenchel transform, $D^*(F_i)$, a function of the generalized forces associated with the internal variables, twice differentiable, non-negative and homogeneous functions of degree two of their respective variables. The evolution laws of the internal variables can be found from the dissipation potentials:

$$\frac{dV_i}{dt} = \frac{\partial \mathcal{D}^*}{\partial (F_i)} \qquad [2.4]$$

Generally, quadratic forms are chosen for both the state potential and the dissipation potential. For example, the state potential is often convex with respect to state variables and internal variables, and concave with respect to temperature: in this case, the conditions of thermodynamic stability are verified *a priori*. It can also be shown that if the dissipation potential is quadratic, positive, null at origin, continuous and convex with respect to its variables, the second principle of thermodynamics is verified *a priori*.

The terms linked to the internal variables appearing in the thermodynamic potential can also be in a "non-quadratic" form (e.g. the terms associated with the plasticity of the material): in this case, a thorough analysis of the stability conditions is necessary to ensure that the principles of thermodynamics are always respected during the evolution of internal variables [MAU 99].

The basic relations of kinematics and statics of continuous media can be found in several works (see, e.g. [GUR 10]) and are briefly recalled here. A material body, B, in the three-dimensional Euclidean space, $\mathcal{E}^3$, consists of material elements and occupies the position $\mathcal{B}_0$ at time $t = 0$ (reference configuration) and $\mathcal{B}$ at time t (actual configuration). Point P belonging to B is denoted by the vector $\mathbf{X}$ in $\mathcal{B}_0$ and by $\mathbf{x}$ in $\mathcal{B}$. A variable (scalar, vector, tensorial, etc.) is defined as material if it is associated with $\mathcal{B}_0$, and spatial if it is associated with $\mathcal{B}$. In addition, a field (vectorial, tensorial, etc.) is represented in a material description (Lagrangian), if it is expressed as a function of the variable $\mathbf{X}$, and in a spatial description (Eulerian) if it is a function of $\mathbf{x}$.

In this context, the change in position, $\mathbf{x}(\mathbf{X}, t)$, the displacement vector, $\mathbf{u}(\mathbf{X}, t)$, and the strain gradient tensor, $\mathbf{F}(\mathbf{X}, t)$, are given, respectively, by:

$$\mathbf{x} = \chi(\mathbf{X}, t) \qquad [2.5]$$

$$\mathbf{u} = \mathbf{x} - \mathbf{X}$$

$$\mathbf{F} = \nabla\chi(\mathbf{X}, t)$$

whereas the Jacobian volume of $\mathbf{F}(\mathbf{X}, t)$, denoted by $J(\mathbf{X}, t)$, is expressed by:

$$J(\mathbf{X}, t) = \det \mathbf{F} > 0 \qquad [2.6]$$

A measurement of the strain, commonly used in the reference configuration, is expressed by:

$$\mathbf{C} = \mathbf{F}^{\mathrm{T}}\mathbf{F} \qquad [2.7]$$

which represents the right Cauchy–Green tensor. The Green–Saint Venant tensor, denoted by $\mathbf{E}$, is linked to $\mathbf{C}$ by the following relation:

$$\mathbf{E} = \frac{1}{2}(\mathbf{C} - \mathbf{I}) = \frac{1}{2}\left(\mathbf{F}^{\mathrm{T}}\mathbf{F} - \mathbf{I}\right) = \frac{1}{2}\left(\mathbf{H} + \mathbf{H}^{\mathrm{T}} + \mathbf{H}^{\mathrm{T}}\mathbf{H}\right) \qquad [2.8]$$

where $\mathbf{H} = \nabla\mathbf{u}$, the displacement gradient.

The static variable, associated with the Green–Saint Venant strain tensor $\mathbf{E}$, is the second Piola tensor, $\mathbf{S}$, while the Cauchy stress tensor, denoted by $\mathbf{T}$, is linked to $\mathbf{S}$ by the following relation:

$$\mathbf{T} = J^{-1}\mathbf{F}\mathbf{S}\mathbf{F}^{\mathrm{T}} \qquad [2.9]$$

$\mathbf{S}$ and $\mathbf{T}$ are, respectively, stress measurements in the reference configuration and in the current configuration. It is important to note that [GUR 10]:

– $\mathbf{T}$ is an objective tensor;

– $\mathbf{C}$, $\mathbf{E}$ and $\mathbf{S}$ are invariant tensors;

– $\mathbf{F}$ is neither objective nor invariant.

It is important to note that – in the TIV framework – the balance equation of a variable $k(\mathbf{x}, t)$ at point $\mathbf{x}$, of a continuous medium at time t, has a general form of the type:

$$\frac{\partial k}{\partial t} + \nabla \cdot (k\mathbf{v}) + \nabla \cdot \mathbf{j}_{cond}[k] = \sigma[k] \qquad [2.10]$$

where $\mathbf{v}$ is the velocity, $\mathbf{j}_{cond}[k]$ the conductive flux density, for k exiting a space region *P* about $\mathbf{x}$, and $\sigma[k]$ the source of k per unit time and volume. Sometimes a distinction is made [MUL 73] between volume sources $\sigma_V[k]$ associated with processes internal to *P*, and surface sources $\sigma_S[k]$ associated with processes external to *P*.

Equation [2.10] can be rewritten in the following form:

$$\frac{\partial k}{\partial t} + \nabla \cdot \mathbf{j}[k] = \sigma[k] \qquad [2.11]$$

where $\mathbf{j}[k]$ expresses the sum of the conduction flux, $\mathbf{j}_{cond}[k]$, and the convection flux, $\mathbf{j}_{conv}[k] = k\mathbf{v}$.

2.2. Development of an ageing-dependent behavior law for organic polymers

The ingredients derived from the experiments described in Chapter 1 and used for modeling the ageing-dependent behavior of organic polymer resins are mentioned briefly below:

– linear viscoelastic behavior, associated with the "opening" of the loading–unloading curves (short-term behavior, see Figures 1.24, 1.32 and 1.33) and with indentation print relaxation phenomena (long-term behavior, see Figures 1.31, 1.34 and 1.35);

– softening behavior, associated with the decrease in the EIT modulus, as a function of the applied load (see Figure 1.29) along the loading/unloading paths (Figure 1.28). This behavior is typical of rubbery and semi-crystalline materials on a macroscopic scale, and is sometimes associated with the Mullins effect [DES 01];

– ageing, characterized by the phenomenological variable γ.

For each of these mechanisms, internal variables are introduced:

– concerning the linear viscoelastic behavior, within the framework of thermodynamics of irreversible processes, the system is supposed to exist in a state of equilibrium (relaxed state). When the system is moved from its equilibrium position through an external perturbation, it returns to equilibrium via a series of elementary mechanisms (macromolecular chain motions), which are represented by j generalized internal variables z_j. A characteristic time τ_j is associated with each internal variable. In accordance with the strain and stress decomposition in spherical and "deviatoric" parts, we introduce two internal variables: a scalar internal variable, associated with the spherical part of the strain tensor, $z^{tr\mathbf{E}_j}$, and another internal tensorial variable, associated with the deviatoric part of the strain tensor, $\mathbf{Z}^{\mathbf{E}_{d_j}}$. We will assume that each internal variable has an equilibrium value defined only from the value taken by the corresponding state variable. Thus, the variations in the internal variables at equilibrium ($z_\infty^{tr\mathbf{E}_j}$ and $\mathbf{Z}_\infty^{\mathbf{E}_{d_j}}$) are correlated with variations in state variables using generalized rigidities ($\beta^{tr\mathbf{E}_j}$ and $\beta^{\mathbf{E}_{d_j}}$), through the relation:

$$dz_\infty^{tr\mathbf{E}_j} = \beta^{tr\mathbf{E}_j}\, d(tr\mathbf{E}) \qquad [2.12]$$

$$d\mathbf{Z}_\infty^{\mathbf{E}_{d_j}} = \beta^{\mathbf{E}_{d_j}}\, d(\mathbf{E}_d)$$

– concerning the softening behavior – phenomenologically similar to the Mullins effect in macroscopic elastomers [DES 01] – an internal variable α affecting the moduli is introduced. This internal variable describes in an approximate manner all the softening effects occurring at small scales (nanoscopic scales or lower), not explicitly observed and associated with the indentation (bending/buckling/damage of the molecular chains between the cross-linkages);

– γ is chosen as the internal variable, given its relevance as an oxidation tracer at the local scale.

Identifying a phenomenological law of evolution of γ, a function of t and z, although limited to given ageing conditions, is a useful step to reproduce numerically the oxidation gradients. To describe this law, a differential equation of the type:

$$\frac{\partial \gamma}{\partial t} = \kappa \frac{\partial^2 \gamma}{\partial z^2} + R(\gamma) \qquad [2.13]$$

inspired by the diffusive–reactive nature of the phenomenon can be proposed. In this equation, κ and $R(\gamma)$ are, respectively, a term analogous to the diffusivity of the species and a source term, associated with the chemical reactions taking place during the oxidation and depending on the oxygen pressure and the temperature. In accordance with the TIV formalism, the structure of equation [2.13] has the same form as general equations [2.10] and [2.11]. We have to express the relation between $R(\gamma)$ and γ by a Gaussian law (see equation [2.14]), which ensures that the input of the "source" is zero for $\gamma = 0$ (non-oxidized polymer), but also for high values of γ associated with the saturation of the oxidation phenomenon. In fact, Colin and Verdu [COL 05] have recently provided a physicochemical interpretation of the development of a saturated zone in the oxidized layer, showing that it forms when the oxidation process takes place in the presence of an oxygen concentration above a certain critical level: oxidation regime "in excess of oxygen".

We thus obtain:

$$R(\gamma) = \tau \,\text{probGauss}(\bar{\gamma}, \sigma_\gamma) \quad [2.14]$$

The term $\text{probGauss}(\bar{\gamma}, \sigma_\gamma)$ expresses the Gaussian probability, defined by an average ($\bar{\gamma}$) and a standard deviation (σ_γ). The parameter τ (in h^{-1}) is introduced to account for the temporal aspect of the phenomenon. The use of a Gaussian law allows us to qualitatively account for a mode of evolution – initiation, propagation and saturation – similar to that of the oxidation reaction [COL 05].

To solve equation [2.13], initial conditions must be specified, i.e.

$$\gamma = 0, \text{ for } t = 0, \forall z \quad [2.15]$$

and the initial conditions at $z = 0$. The boundary condition ($z = 0$) is given experimentally by the values of γ observed near to the sample surface (see Figure 1.19 in Chapter 1), denoted as γ_{max}, analytically expressed by equation [1.4] (see Chapter 1).

Equation [2.13] is solved using a MATLAB® script using the "ode45" method, based on an explicit Runge–Kutta algorithm.

The four parameters of the equation are identified from the experimental curves of γ and are presented in Table 2.1.

	Air	**2 bar O_2**	**2 bar O_2/ Air**
κ [$\mu m^2/h$]	18	130	7.22
τ [h^{-1}]	0.00038	0.0028	7.37
$\bar{\gamma}$ [-]	0.5	0.5	1
σ_γ [-]	0.1	0.1	1

Table 2.1. *Values of the parameters of the evolutive law of γ for ageing in air at atmospheric pressure or under 2 bar O_2. (see also [GIG 16a, MIN 13, MIN 14])*

The numerical profiles obtained from these parameters are compared with the experimental points in Figure 2.1, for conditioning under 2 bar O_2, and in Figure 2.2, in atmospheric air.

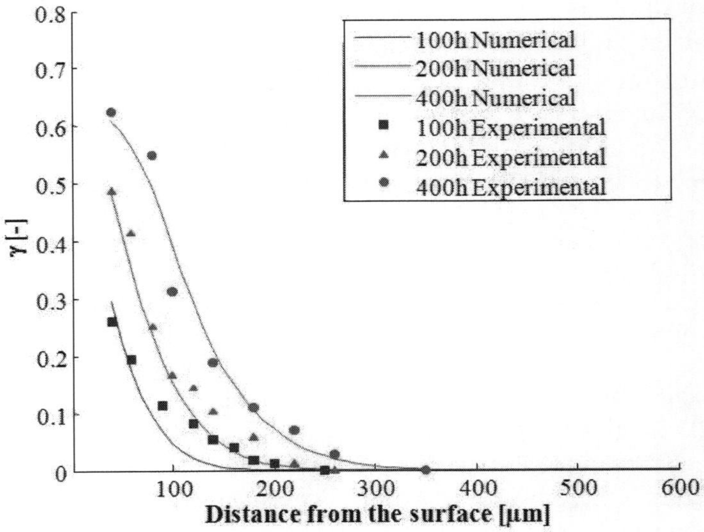

Figure 2.1. *Numerical and experimental profiles of γ for the TACTIX resin aged under 2 bar O_2 at 150°C (see also [GIG 16a, MIN 13, MIN 14]). For a color version of this figure, see www.iste.co.uk/gigliotti/mechanical.zip*

It can be seen that the numerical profiles are in agreement with the experimental measurements, except in the case of ageing of 168 hours under 2 bar O_2.

In the remainder of this chapter, this evolutive law will therefore be used only to study ageing under 2 bar O_2 with a duration of less than 72 hours.

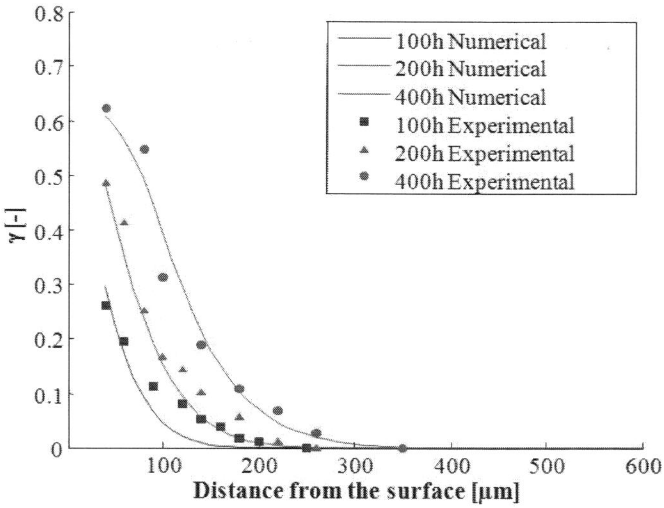

Figure 2.2. *Numerical and experimental profiles of γ for the TACTIX resin aged in air at atmospheric pressure at 150°C (see also [GIG 16a, MIN 13, MIN 14]). For a color version of this figure, see www.iste.co.uk/gigliotti/mechanical.zip*

In Table 2.1, the values of the parameters κ and τ depend on the partial pressure of oxygen; a significant decrease is observed when we pass from ageing under 2 bar O_2 to ageing in air at atmospheric pressure. It is also noted that the ratios of these two parameters are very close to 7.3. In addition, the parameters $\bar{\gamma}$ and σ_γ, which characterize the Gaussian law, seem to not be influenced by the partial pressure of oxygen. This observation can be directly compared with the measurements of γ and γ_{max} presented in Chapter 1, which had the same acceleration factors close to 7.

Equation [2.13] – complemented by the form of the source term, equation [2.14] – solved with the appropriate initial (see equation [2.15]) and boundary conditions (see equation [1.4], in Chapter 1) allows us to predict the gradients associated with thermo-oxidation phenomena, within aged material. The existence of a numerical solution to the differential equation [2.13] will allow the local behavior law of the ageing-dependent polymer to be numerically parameterized, as a function of γ. These developments will be presented in Chapter 3.

To develop the local behavior law, the ageing-dependent polymer, the Helmholtz specific energy, ψ (state potential), is taken quadratically with respect to the state variables and the internal variables, but concave with respect to temperature. Under isothermal conditions, the temperature dependence will not be explicitly considered:

$$\psi_R\left(\mathrm{tr}\mathbf{E}, \mathbf{E}_d, z^{\mathrm{tr}\mathbf{E}_j}, \mathbf{Z}^{\mathbf{E}_{d_j}}, \alpha, \gamma\right) \qquad [2.16]$$

$$\begin{aligned}\rho_R\psi_R = &\frac{3}{2}K_\infty^{\gamma*}(\alpha)(\mathrm{tr}\mathbf{E})^2 + G_\infty^{\gamma*}(\alpha)|\mathbf{E}_d|^2\\ &+\frac{3}{2}\sum_{j=1}^{n} K_j^{\gamma*}(\alpha)\left(z^{\mathrm{tr}\mathbf{E}_j} - z_\infty^{\mathrm{tr}\mathbf{E}_j}\right)^2\\ &+\sum_{j=1}^{m} G_j^{\gamma*}(\alpha)\left(\mathbf{Z}^{\mathbf{E}_{d_j}} - \mathbf{Z}_\infty^{\mathbf{E}_{d_j}}\right)^2\end{aligned}$$

where $K_\infty^{\gamma*}(\alpha)$, $G_\infty^{\gamma*}(\alpha)$, $z_\infty^{\mathrm{tr}\mathbf{E}_j}$, $\mathbf{Z}_\infty^{\mathbf{E}_{d_j}}$ are, respectively, the relaxed values of the elasticity and shear modulus – dependent on the softening variable α, ageing γ and the internal variables. $K_j^{\gamma*}(\zeta)$ and $G_j^{\gamma*}(\zeta)$ are the generalized moduli related to viscoelastic fluctuations close to the equilibrium position, which are also dependent on the softening variable α and ageing.

The dependence of the moduli of the softening variable α is of the type:

$$K_\infty^{\gamma*}(\alpha) = (1-\alpha)\,K_\infty^{\gamma} \qquad [2.17]$$

$$G_\infty^{\gamma*}(\alpha) = (1-\alpha)\,G_\infty^{\gamma}$$

$$K_j^{\gamma*}(\alpha) = (1 - \alpha)\, K_j^{\gamma}$$

$$G_j^{\gamma*}(\alpha) = (1 - \alpha)\, G_j^{\gamma}$$

The state laws can be calculated from the thermodynamic potential:

$$\mathrm{tr}\mathbf{S} = \rho_R \frac{\partial \psi_R}{\partial(\mathrm{tr}\mathbf{E})} = 3K_\infty^{\gamma*}(\alpha)(tr\boldsymbol{E}) - 3\sum_{j=1}^{n} K_j^{\gamma*}(\alpha)\, \beta^{\mathrm{tr}\mathbf{E}_j}(z^{\mathrm{tr}\mathbf{E}_j} - z_\infty^{\mathrm{tr}\mathbf{E}_j}) \quad [2.18]$$

$$\mathbf{S}_d = \rho_R \frac{\partial \psi_R}{\partial(\mathbf{E}_d)} = 2G_\infty^{\gamma*}(\alpha)\mathbf{E}_d - 2\sum_{j=1}^{m} G_j^{\gamma*}(\alpha)\, \beta^{\mathbf{E}_{d_j}}(\mathbf{Z}^{\mathbf{E}_{d_j}} - \mathbf{Z}_\infty^{\mathbf{E}_{d_j}})$$

$$A^{\mathrm{tr}\mathbf{E}_j} = -\rho_R \frac{\partial \psi_R}{\partial\left(z^{\mathrm{tr}\mathbf{E}_j}\right)} = -3K_j^{\gamma*}(\alpha)\,(z^{\mathrm{tr}\mathbf{E}_j} - z_\infty^{\mathrm{tr}\mathbf{E}_j})$$

$$\mathbf{A}^{\mathbf{E}_{d_j}} = -\rho_R \frac{\partial \psi_R}{\partial\left(\mathbf{Z}^{\mathbf{E}_{d_j}}\right)} = -2G_j^{\gamma*}(\alpha)\,(\mathbf{Z}^{\mathbf{E}_{d_j}} - \mathbf{Z}_\infty^{\mathbf{E}_{d_j}})$$

$$A^{\alpha} = -\rho_R \frac{\partial \psi_R}{\partial(\alpha)} = \frac{3}{2}K_\infty^{\gamma}(\mathrm{tr}\mathbf{E})^2 + G_\infty^{\gamma}|\mathbf{E}_d|^2 + \frac{3}{2}\sum_{j=1}^{n} K_j^{\gamma}\,(z^{\mathrm{tr}\mathbf{E}_j} - z_\infty^{\mathrm{tr}\mathbf{E}_j})^2 + \sum_{j=1}^{m} G_j^{\gamma}(\mathbf{Z}^{\mathbf{E}_{d_j}} - \mathbf{Z}_\infty^{\mathbf{E}_{d_j}})^2$$

$$A^{\gamma} = -\rho_R \frac{\partial \psi_R}{\partial(\gamma)} = \frac{3}{2}(\partial K_\infty^{\gamma*}/\,\partial\gamma)(\mathrm{tr}\mathbf{E})^2 + (\partial G_\infty^{\gamma*}/\,\partial\gamma)|\mathbf{E}_d|^2$$

$$+\frac{3}{2}\sum_{j=1}^{n}(\partial K_j^{\gamma*}/\,\partial\gamma)\,(z^{\mathrm{tr}\mathbf{E}_j} - z_\infty^{\mathrm{tr}\mathbf{E}_j})^2 + \sum_{j=1}^{m}(\partial G_j^{\gamma*}/\,\partial\gamma)(\mathbf{Z}^{\mathbf{E}_{d_j}} - \mathbf{Z}_\infty^{\mathbf{E}_{d_j}})^2$$

$\mathrm{tr}\mathbf{S}$ and $\mathbf{S}_d$ represent the “elastic” stress, $A^{\mathrm{tr}\mathbf{E}_j}$ and $\mathbf{A}^{\mathbf{E}_{d_j}}$ the “viscoelastic” stress, and A^{α} and A^{γ} the “thermodynamic forces” associated with the respective internal variables, α and γ. We note that the thermodynamic force A^{α}, associated with the internal variable α, is always positive.

The thermodynamic dissipation Φ, expressed as:

$$\Phi = A^{\gamma}\frac{d\gamma}{dt} + A^{\alpha}\frac{d\alpha}{dt} + \sum_{j=1}^{n} A^{\mathrm{tr}\mathbf{E}_j}\frac{dz^{\mathrm{tr}\mathbf{E}_j}}{dt} + \sum_{j=1}^{n}\mathbf{A}^{\mathbf{E}_{d_j}} : \frac{d\mathbf{Z}^{\mathbf{E}_{d_j}}}{dt} \geq 0 \qquad [2.19]$$

must be positive (or zero) in order for the second principle of thermodynamics to be satisfied. The dissipation can be divided into three terms: one (Φ_γ) linked to dissipation by thermo-oxidative ageing (γ), another (Φ_α) associated with softening phenomena (α) and a third (Φ_{visco}) associated with the viscoelastic behavior ($z^{tr\mathbf{E}_j}$ and $\mathbf{Z}^{\mathbf{E}_{d_j}}$), expressed, respectively, by:

$$\Phi_\gamma = A^\gamma \frac{d\gamma}{dt} \geq 0 \qquad [2.20]$$

$$\Phi_\alpha = A^\alpha \frac{d\alpha}{dt} \geq 0$$

$$\Phi_{visco} = \sum_{j=1}^{n} A^{tr\mathbf{E}_j} \frac{dz^{tr\mathbf{E}_j}}{dt} + \sum_{j=1}^{n} \mathbf{A}^{\mathbf{E}_{d_j}} : \frac{d\mathbf{Z}^{\mathbf{E}_{d_j}}}{dt} \geq 0$$

When decoupling dissipation phenomena, the three terms being “non-negative” are sufficient conditions for the second principle of thermodynamics to be respected.

As for the first dissipation term (Φ_γ), the form of the phenomenological law of γ (equation [2.13]) leads to the conclusion that the internal ageing variable increases with time monotonically and that the condition:

$$\frac{d\gamma}{dt} \geq 0 \qquad [2.21]$$

is respected. The non-negativity of A^γ must be verified *a posteriori*. As for the dissipation term associated with softening phenomena (Φ_α), the thermodynamic force A^α is always non-negative, since in equation [2.18], the polymer moduli are always positive, and all state and internal variables are quadratic. According to equation [2.20], the condition:

$$\frac{d\alpha}{dt} \geq 0 \qquad [2.22]$$

must be respected. Moreover, equation [2.17] – linking the elasticity moduli of the polymer with the internal softening variable – forces the variable α to be between 0 and a maximum value α_{max}, less than 1. The values 0 and α_{max} therefore correspond, respectively, to the state of the polymer before the application of the load and to a limit softened state, in which the molecular structure of the polymer has evolved to adapt to the applied load. For the softening behavior, an explicit functional relation between α and A^{α} is given. Moreover, in order to limit the softening variable within the indicated range of values, while simultaneously minimizing the number of coefficients required, an exponential correlation in the form:

$$\alpha = b(1 - \exp^{(-c\langle A^{\alpha}\rangle)}) \quad [2.23]$$

is given, in which:

$$\langle A^{\alpha}\rangle = A^{\alpha} \textbf{ if } A^{\alpha} \geq Max_t A^{\alpha} \quad [2.24]$$

$$\langle A^{\alpha}\rangle = Max_t A^{\alpha} \textbf{ if } A^{\alpha} \leq Max_t A^{\alpha}$$

$Max_t A^{\alpha}$ being the maximum value reached by A^{α} in the time interval between 0 and t. In equation [2.23], b and c are both phenomenological coefficients to be identified, subject to the constraints:

$$c > 0 \quad [2.25]$$

$$b \leq 1$$

For viscoelastic dissipation (Φ_{visco}), a dissipation potential:

$$\mathcal{D}^*(A^{trE_j}, \mathbf{A}^{\mathbf{E}_{d_j}}) \quad [2.26]$$

in the form:

$$\mathcal{D}^* = \frac{1}{2}\sum_{j=1}^{n} B_{tr\mathbf{E}_j} \left(\rho^{-1}A^{tr\mathbf{E}_j}\right)^2 + \frac{1}{2}\sum_{j=1}^{n}\left(\rho^{-1}\mathbf{A}^{\mathbf{E}_{d_j}}\right) : \mathbb{B}_{\mathbf{E}_{d_j}} \left(\rho^{-1}\mathbf{A}^{\mathbf{E}_{d_j}}\right) \quad [2.27]$$

can be introduced, in which $B_{tr\mathbf{E}_j}$ and $\mathbb{B}_{\mathbf{E}_{d_j}}$ are phenomenological coefficients associated with the relaxation phenomena. The state equation, equation [2.18], can be combined with the dissipation potential (see equation [2.27]) to obtain the evolution laws of the internal variables, $z^{tr\mathbf{E}_j}$ and $\mathbf{Z}^{\mathbf{E}_{d_j}}$, towards the equilibrium values, $z_\infty^{tr\mathbf{E}_j}$ and $\mathbf{Z}_\infty^{\mathbf{E}_{d_j}}$, following "relaxation" equations of the type:

$$\frac{dz^{tr\mathbf{E}_j}}{dt} = \frac{\partial \mathcal{D}^*}{\partial\left(\rho A^{tr\mathbf{E}_j}\right)} = -\left(\tau^{tr\mathbf{E}_j}\right)^{-1}\left(z^{tr\mathbf{E}_j} - z_\infty^{tr\mathbf{E}_j}\right) \quad [2.28]$$

$$\frac{d\mathbf{Z}^{\mathbf{E}_{d_j}}}{dt} = \frac{\partial \mathcal{D}^*}{\partial\left(\rho \mathbf{A}^{\mathbf{E}_{d_j}}\right)} = -\left(\tau^{\mathbf{E}_{d_j}}\right)^{-1}\left(\mathbf{Z}^{\mathbf{E}_{d_j}} - \mathbf{Z}_\infty^{\mathbf{E}_{d_j}}\right)$$

where $\tau^{tr\mathbf{E}_j}$ and $\tau^{\mathbf{E}_{d_j}}$ are characteristic relaxation times associated with relaxation phenomena. It should be noted that, with the definitions of equations [2.26] to [2.28], the Clausius–Duhem inequality is exactly respected for viscoelastic behavior.

From equations [2.12], [2.18] and [2.28], it is possible to introduce the "instantaneous glassy" modulus of the polymer, defined by:

$$K_0^{\gamma*}(\alpha) = (1-\alpha)\left(K_\infty^\gamma + \sum_{j=1}^{n} K_j^\gamma(\alpha)\left(\beta^{tr E_j}\right)^2\right) = (1-\alpha)K_0^\gamma \quad [2.29]$$

$$G_0^{\gamma*}(\alpha) = (1-\alpha)\left(G_\infty^\gamma + \sum_{j=1}^{n} G_j^\gamma(\alpha)\left(\beta^{tr E_j}\right)^2\right) = (1-\alpha)G_0^\gamma$$

The state equations, defining the elastic stress, therefore become the following:

$$\mathrm{tr}\mathbf{S} = 3K_0^{\gamma*}(\alpha)(tr\boldsymbol{E}) - 3\sum_{j=1}^{n} K_j^{\gamma*}(\alpha)\ \beta^{\mathrm{tr}\mathbf{E}_j} z^{\mathrm{tr}\mathbf{E}_j} \quad [2.30]$$

$$\mathbf{S}_d = 2G_0^{\gamma*}(\alpha)\mathbf{E}_d - 2\sum_{j=1}^{m} G_j^{\gamma*}(\alpha)\ \beta^{\mathbf{E}_{d_j}} \mathbf{Z}^{\mathbf{E}_{d_j}}$$

Finally, the parameters to be identified are: K_0, G_0, b, c and $\forall$ j (K_j, G_j, $\beta^{\mathrm{tr}\mathbf{E}_j}$, $\beta^{\mathbf{E}_{d_j}}$, $\tau^{\mathrm{tr}\mathbf{E}_j}$ and $\tau^{\mathbf{E}_{d_j}}$). Moreover, we assume that $K_j = G_j$, $\tau_j = \tau^{\mathrm{tr}\mathbf{E}_j} = \tau^{\mathbf{E}_{d_j}}$, $n = m$, $\beta^{\mathrm{tr}\mathbf{E}_j} = 1$, $\beta^{\mathbf{E}_{d_j}} = 1$.

According to Cunat [CUN 01], an appropriate number of relaxation times must be taken into account, in order to best represent the response of the material. In addition, at each characteristic relaxation time, there must be an associated relaxed bulk modulus and a relaxed shear modulus. In order to reduce the number of variables, a particular form can be given to the relaxation time spectrum, i.e. the "relaxed moduli versus characteristic relaxation time" curve. In this work, the modeling of this curve is based on the double-box distribution proposed by Kovacs and illustrated in Figure 2.3(a): it is a type of modeling used in the study of the physical ageing of polymers [CUN 01].

According to this empirical spectrum, two different blocks (block 1 and block 2 in Figure 2.3(a)) are used to separately model the relaxation phenomena in the short and long terms. The dimensions of each block (height and width) give the relative weights of the relaxation phenomena, both in the short and in the long term, and their impact on the relaxation spectrum. The distribution of the spectrum over the relaxation times can be changed using larger (and shorter) blocks – in this case, the relaxation phenomena are distributed over several decades – or narrower (and taller) blocks, leading to significant relaxation phenomena, only active over a few decades. The total area of the double-box distribution must always be equal to 1.

The double-box distribution has therefore been improved in order to model the relaxation phenomena that occur in the short and long terms. A total of 33 characteristic relaxation times – from 1 second up to 100,000 seconds (the approximate total duration of the test) in a logarithmic scale – were independently selected and grouped into two blocks (see Figure 2.3(b)). The first block, corresponding to short relaxation times, has been modeled using a constant function whose main parameters are the dimensionless relaxed modulus value (P_1) and the relaxation time limit beyond which the function is zero (P_2). On the other hand, for long relaxation times, the second block was modeled using a Gaussian distribution – defined by an average (P_3) and a standard deviation (P_4) – multiplied by an intensity correction parameter (P_5), to correct the total area of the Gaussian distribution. Since the sum of the dimensionless relaxation moduli must be equal to 1, the parameter P_5 is not independent.

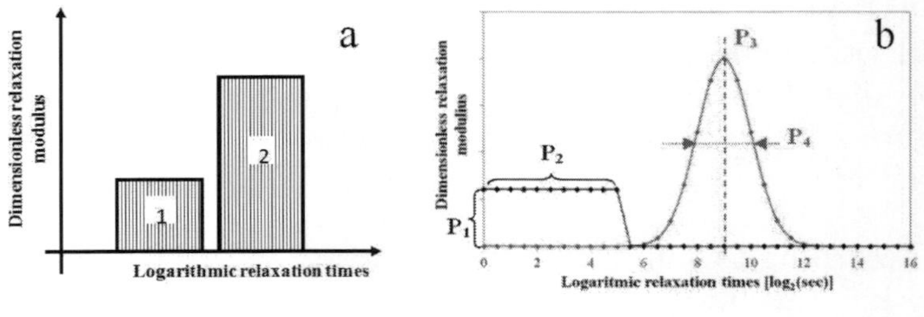

Figure 2.3. *a) Kovacs' empirical double-box distribution. b) Adaptation of Kovacs' double-box distribution and a series of relaxation parameters (see also [GIG 16a, MIN 13, MIN 14]). For a color version of this figure, see www.iste.co.uk/gigliotti/mechanical.zip*

Four variables ($P_1 – P_4$) are therefore sufficient to express the viscoelastic behavior of the polymer. By adding the two parameters (K_0 and G_0) describing the elastic behavior and the two parameters (b and c) describing the softening phenomena, the total number of variables of the viscoelastic model amounts to eight.

2.3. Taking into account the initial inelastic and chemical strains

Strains of thermal and chemical origin, the latter associated with thermo-oxidation phenomena (see the phenomenological description in the introduction and Chapter 1), represent irreversible inelastic strains within the polymer and the OMC and will be introduced into the thermodynamic formulation. More specifically:

– in the initial state, the initial inelastic strain ($\mathbf{E}^{in}$), producing the matrix shrinkage observed on the surface of the virgin composite, which simultaneously takes into account the matrix strain by thermal expansion ($\mathbf{E}^{T}$) and that generated by polishing the surface;

– in the oxidized state, chemical inelastic strain ($\mathbf{E}^{ch}(\gamma)$), developing during oxidation, mainly due to the release of volatile products and water molecules.

These inelastic strains induce a variation in the volume of the matrix and they therefore must be introduced into the spherical part of the thermodynamic potential of the polymer matrix.

The thermodynamic potential therefore becomes:

$$\begin{aligned}\rho_R\psi_R = &\frac{3}{2}K_\infty^{\gamma*}(\alpha)(\mathrm{tr}\mathbf{E})^2 - 3K_\infty^{\gamma*}(\alpha)\mathrm{tr}\mathbf{E}^{in}\mathrm{tr}\mathbf{E} - 3K_\infty^{\gamma*}(\alpha)\mathrm{tr}\mathbf{E}^{ch}(\gamma)\mathrm{tr}\mathbf{E} \\ &+G_\infty^{\gamma*}(\alpha)|\mathbf{E}_d|^2 + \frac{3}{2}\sum_{j=1}^{n} K_j^{\gamma*}(\alpha)\left(z^{\mathrm{tr}\mathbf{E}_j} - z_\infty^{\mathrm{tr}\mathbf{E}_j}\right)^2 \\ &+\sum_{j=1}^{m} G_j^{\gamma*}(\alpha)\left(\mathbf{Z}^{\mathbf{E}_{d_j}} - \mathbf{Z}_\infty^{\mathbf{E}_{d_j}}\right)^2 \qquad [2.31]\end{aligned}$$

From this new expression of the thermodynamic potential, the new state laws taking into account the initial inelastic strains and those of chemical origin, can be derived. This completes the modeling of the local mechanical behavior dependent on the ageing of thermo-oxidized polymers. The next chapter (Chapter 3) will be devoted to the identification, use for simulation, and validation of these models. To conclude, it should be noted that alternative approaches, in particular to modeling mechanical behavior dependent on the thermo-oxidative ageing of elastomers, can be found in [JOH 13, JOH 14].

3

Identification and Simulations

This chapter is devoted to the identification of the mechanical behavior law of virgin and oxidized polymers and the chemical strains associated with thermo-oxidation. This identification is carried out by combining the experimental measurements presented in Chapter 1 and the model developed in Chapter 2. Several comparisons between the results from numerical simulations and experimental tests validate the approach and the model, both for the polymer and for the unidirectional composite.

3.1. Identifying the behavior law of thermo-oxidized polymers through the inverse analysis of ultra-micro-indentation tests

3.1.1. *The method to identify the local mechanical behavior of virgin and oxidized polymers*

A 3D numerical model of the UMI test was developed in the finite element commercial code ABAQUS© [DAS 09]. The model consists of two parts: the indenter and a polymer sample. In order to get closer to the actual conditions of the indentation test, the diamond tip was drawn by 3D mapping of the indentation apparatus (see Figure 3.1(a)). The indenter is assumed to be perfectly rigid and a reference point (the tip of the indenter) is defined: all displacements and all reactions are calculated by ABAQUS at this point.

Modeling the exact geometry of the indenter allows for the creation of a 3D shape of the indentation print that can be compared with IM measurements, but prevents the identification of planes of symmetry. This requires the polymer to be modeled in its entirety, as shown in Figure 3.1(b); for the mesh 32,000 cubic elements with reduced integration of type C3D8R are used (see Figure 3.1(c)).

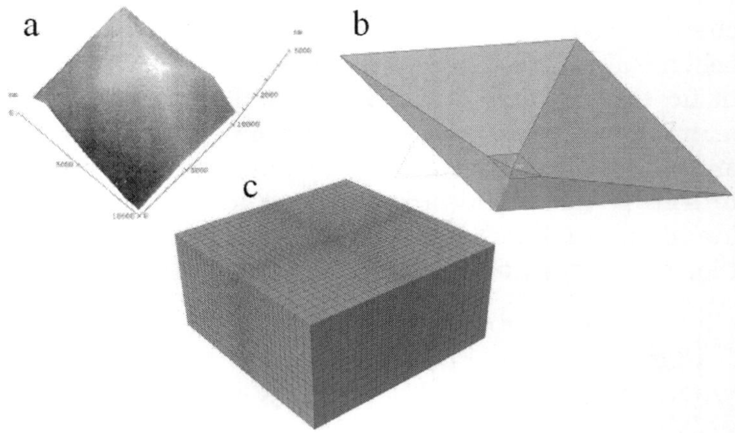

Figure 3.1. *a) Mapping of the diamond indenter. b) Geometry of the indenter reconstituted in ABAQUS. c) ABAQUS model of the polymer tested by UMI (see also [GIG 16a, MIN 13, MIN 14]). For a color version of this figure, see www.iste.co.uk/gigliotti/mechanical.zip*

In accordance with [KER 08, BRE 09, RAU 10], no friction was considered between the surface of the indenter and that of the sample: this approximation may be questionable, but we must bear in mind that an experimental value for the friction coefficient is difficult to identify. Quasi-static simulations, using the VISCO procedure, with the NLGEOM (*non-linear geometry*) option to account for large strains, were carried out. Finally, the constitutive law was introduced via a USDFLD subroutine, written in FORTRAN language.

The behavior law parameters were optimized using the open source software DAKOTA© [SAN 09]. This software has a large library of identification algorithms: several methods are available, based on the gradient calculation, on genetic techniques, on the random generation of solutions, and so on.

The constitutive law, whose parameters are to be optimized, is rather complex. Several parameters are to be identified (see Chapter 2) and the solution can easily have several local minima. This is why we have chosen to follow an identification protocol composed of two optimization algorithms of a different nature: a genetic algorithm and an algorithm based on the gradient calculation. The genetic algorithm is useful for scanning the domain of possible solutions. Simultaneously, over several generations, it results in very few errors, and thus verifies the presence of multiple solutions. The presence of multiple solutions indicates that the information provided experimentally is not sufficient for an unambiguous interpretation of the polymer behavior and that a more complete experimental database is required. On the contrary, if all the parameter sets proposed by DAKOTA, for the same generation, are similar, the optimal solution is not far away. The final identification is carried out by a Gauss–Newton algorithm which, after a few iterations, identifies the best set of parameters for the desired law of behavior. The "dialogue" between DAKOTA and ABAQUS is schematized in Figure 3.2.

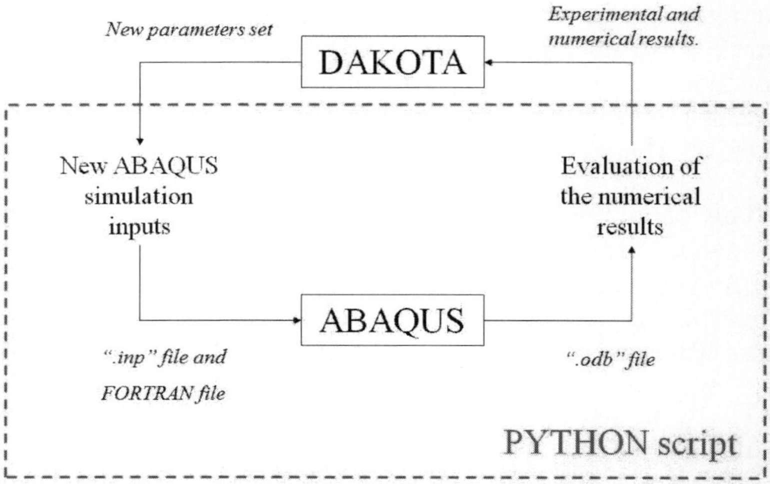

Figure 3.2. *Diagram of the optimization algorithm and the procedure to identify the parameters of the ageing-dependent behavior law of the polymer (see also [GIG 16a, MIN 13, MIN 14]). For a color version of this figure, see www.iste.co.uk/gigliotti/mechanical.zip*

For a chosen optimization algorithm, the data needed to solve the optimization problem are:

– the difference between the numerical and experimental indentation and recovery curves;

– the number and limits of the parameters of the polymer law of behavior.

Following an iteration, DAKOTA uses a new set of parameters for the local behavior law of the polymer.

A PYTHON© [PYT 12] script was developed to manage the data exchange and exploit the result of the numerical simulation, to give DAKOTA the deviation between the numerical and experimental curves. This difference is calculated over fifteen time points, opportunely chosen in order to have a good discretization of the indentation and recovery curves.

3.1.2. *Identification of local mechanical behavior of virgin polymers*

The optimized set of parameters of the law of behavior of the virgin material is summarized in Table 3.1:

K_0	G_0	B	C	P_1	P_2	P_3	P_4
[MPA]	[MPA]	[-]	[MPA^1]	[-]	[log$_2$(sec)]	[log$_2$(sec)]	[log$_2$(sec)]
3,540	1,310	0.765	13,656	0.015	4	8.1	1.8

Table 3.1. *Parameters identified for the local behavior law of the virgin TACTIX polymer (see also [GIG 16a, MIN 13, MIN 14])*

Figure 3.3 presents a comparison between the numerical simulations and the experimental results, showing that the model is able to correctly reproduce both the indentation curve and the calibration recovery curve.

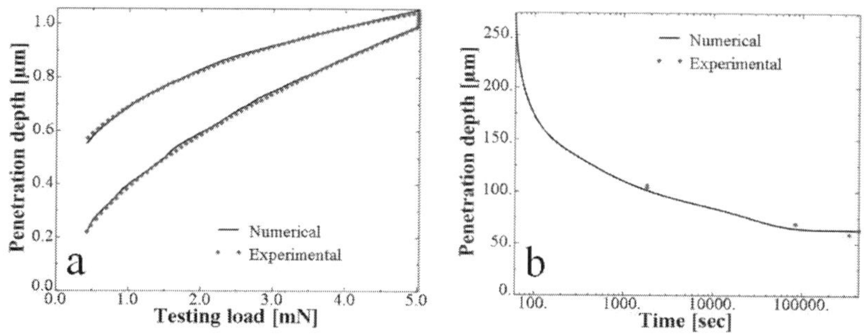

Figure 3.3. *Identification of parameters for the local behavior law of the virgin TACTIX polymer (see also [GIG 16a, MIN 13, MIN 14]). For a color version of this figure, see www.iste.co.uk/gigliotti/mechanical.zip*

In order to validate the constitutive law, other simulations were carried out using the optimized parameter set (see Table 3.1).

In Figure 3.4, the experimental (UMI) and numerical indentation profiles have been plotted on a graph, 10 minutes and 1 day after the indentation tests: the maximal print displacement is well estimated and the numerical model follows the trend and its evolution during recovery.

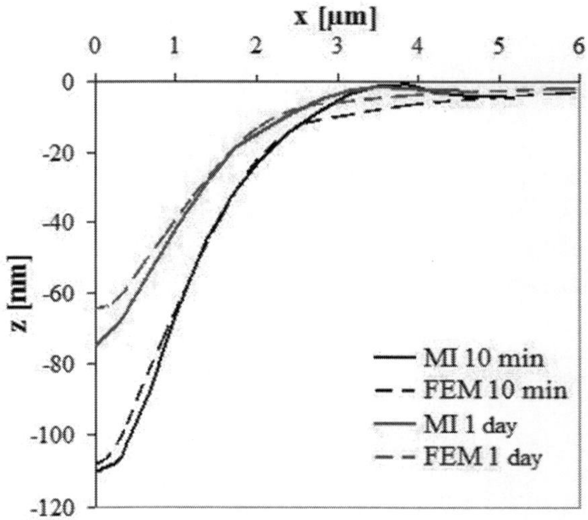

Figure 3.4. *Validation of the local behavior law of virgin TACTIX polymer: indentation profiles after 10 minutes and 1 day of testing (see also [GIG 16a, MIN 13, MIN 14]). For a color version of this figure, see www.iste.co.uk/gigliotti/mechanical.zip*

Figure 3.5 compares the experimental measurements with the simulations for an indentation test without the creep phase and for the subsequent recovery phase.

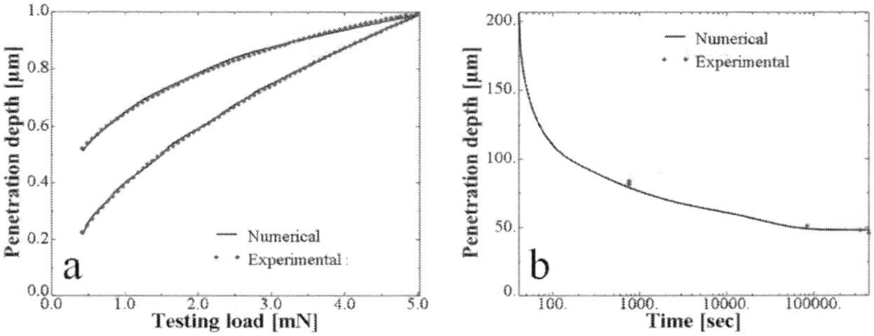

Figure 3.5. *Validation of the behavior law of the virgin TACTIX polymer: indentation a) without the creep phase and b) for the associated recovery phase (see also [GIG 16a, MIN 13, MIN 14]). For a color version of this figure, see www.iste.co.uk/gigliotti/mechanical.zip*

Figure 3.6 presents simulation/indentation comparisons with different loading speeds (a) and different creep times (b):

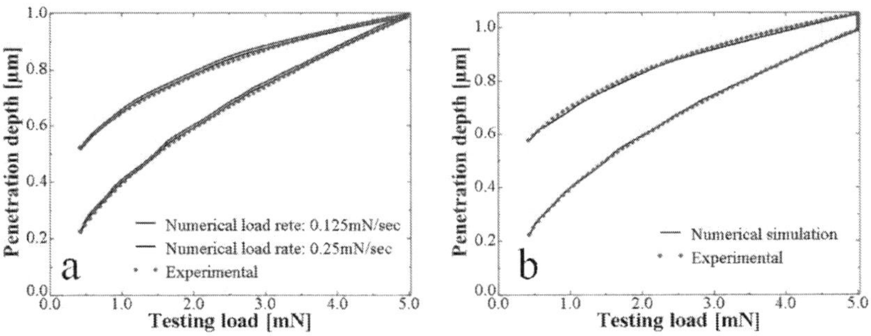

Figure 3.6. *Validation of the behavior law of the virgin TACTIX polymer: indentation tests a) with different loading speeds and b) with a hold stage of 60 seconds (see also [GIG 16a, MIN 13, MIN 14]). For a color version of this figure, see www.iste.co.uk/gigliotti/mechanical.zip*

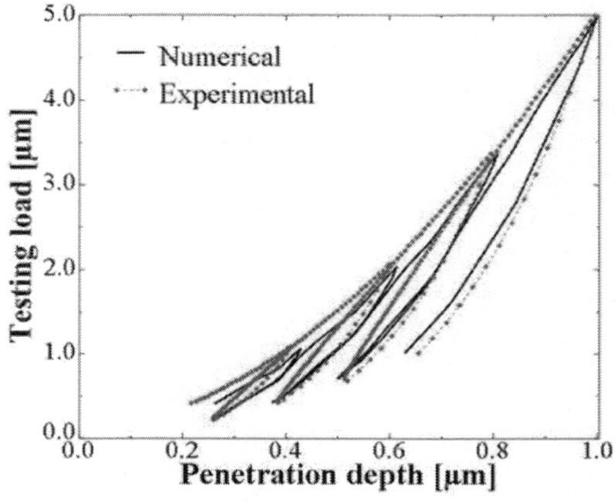

Figure 3.7. *Validation of the behavior law of virgin TACTIX polymer: progressive indentation tests (see also [GIG 16a, MIN 13, MIN 14]). For a color version of this figure, see www.iste.co.uk/gigliotti/mechanical.zip*

Finally, Figure 3.7 shows a progressive indentation test. The numerical model follows the experimental behavior of the polymer. At the maximum load of each loop, the numerical and experimental displacement are almost the same: at the first, the indentation loop calculated is approximately 0.43 μm, instead of 0.41 μm (4% more), at the second, 0.61 μm, instead of 0.60 μm (1% more), at the third, 0.81 μm, instead of 0.80 μm (1% more) and, finally, at the last loop, 1.00 μm, instead of 0.99 μm (1% more).

It is observed, however, that the simulated loop opening is not entirely identical to the experimental one, with the greatest deviation at the first loop (4% error). This suggests that the viscoelastic behavior in the short term should be modeled more accurately: in this case, however, the double box concept for the description of the relaxation time distribution (see Figure 2.3 in Chapter 2) should be abandoned and a more complex relaxation spectrum (for example, supplemented by a third block accounting for behavior in the short term) should be adopted.

The behavior law developed thus allows us to reproduce numerically a wide variety of tests: the mechanical behavior of the polymer on the local scale is reasonably well illustrated and the constitutive law identified proves to be robust and efficient.

3.1.3. *Identifying the local mechanical behavior of the oxidized polymer*

The procedure for identifying the parameters of the behavior law, detailed above for the virgin sample, is carried out for each polymer sample which has undergone thermo-oxidative ageing at 150°C under 2 bar O_2, of between 1 day and 7 days. This procedure makes it possible to obtain a different set of parameters for each value of γ. In Chapter 1 (see Figures 1.21 and 1.22), we saw that, at each value of γ, there is only one corresponding mechanical response of the polymer, irrespective of the history of its ageing under atmospheric air or oxygen pressure, supporting the hypothesis of representing the parameters of the behavior law of the oxidized polymer as a function of the parameter γ.

The evolutions of the eight parameters of the constitutive law as a function of the level of oxidation are presented in the following figures. The very weak scatter of the results in these figures supports the robustness of the optimization procedure and the relevance of the constitutive law chosen.

By analyzing the evolution of the instantaneous glass elasticity modulus (K_0 and G_0) represented as a function of γ (see Figure 3.8), linear correlations are identified.

Indeed, all the points appear to lie around two average curves – black for K_0 and red for G_0 – whose analytical expressions are the following:

$$K_0(\gamma) = 3{,}042\,\gamma + 3{,}540\ [\text{MPa}] \qquad [3.1]$$

$$G_0(\gamma) = 1{,}128\,\gamma + 1{,}310\ [\text{MPa}]$$

From the evolutions of the compressibility and shear moduli, the instantaneous elastic behavior of the polymer can also be expressed in terms of Young's modulus (E_0). Figure 3.9 shows the evolution of the E_0 modulus as a function of the phenomenological ageing parameter, γ. There is a steady linear increase in the E_0 modulus (see equation [3.2]), indicating that the polymer in the oxidized layer is more rigid than that in the virgin state. The linear correlation reasonably derives from the formulation of γ, defined from the indentation modulus (EIT), which is itself an "elastic modulus":

$$E_0(\gamma) = 3{,}011\,\gamma + 3{,}500\ [\text{MPa}] \qquad [3.2]$$

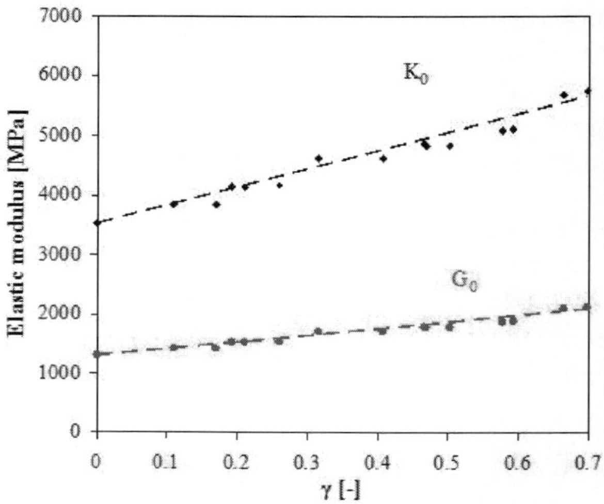

Figure 3.8. *Instantaneous glass elastic modulus (K_0 and G_0), as a function of γ (oxidized TACTIX resin) (see also [GIG 16a, MIN 13, MIN 14]). For a color version of this figure, see www.iste.co.uk/gigliotti/mechanical.zip*

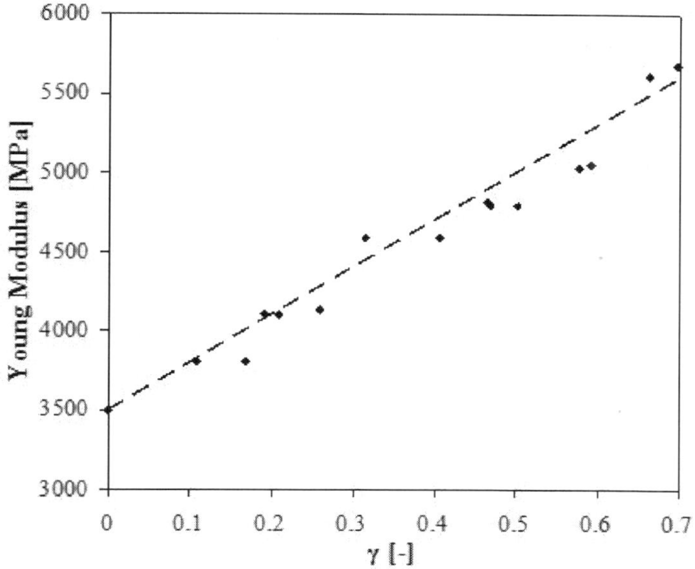

Figure 3.9. *Evolution of Young's modulus E_0 as a function of γ (oxidized TACTIX resin) (see also [GIG 16a, MIN 13, MIN 14]). For a color version of this figure, see www.iste.co.uk/gigliotti/mechanical.zip*

The variations in the softening parameters are shown in Figure 3.10. Parameter b is related to the softening rate of the elastic modulus. Its decrease during ageing leads to a softened modulus on the initial modulus closer to 1 and therefore has, in the mechanical behavior of the oxidized polymer, an increasingly weaker influence during ageing. On the other hand, parameter c affects the "propagation velocity" of the softened zone: high c values correspond to more extensive softer zones in the polymer.

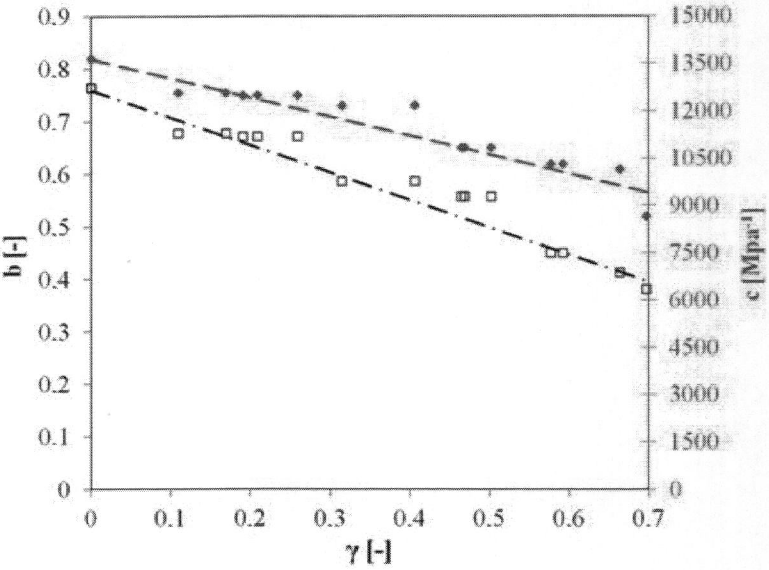

Figure 3.10. *Evolution of softening parameters as a function of γ (oxidized TACTIX resin) (see also [GIG 16a, MIN 13, MIN 14]). For a color version of this figure, see www.iste.co.uk/gigliotti/mechanical.zip*

The softening parameters (b and c) are approximated by linear functions of the coefficient γ and the corresponding expressions are given in equation [3.3]:

$$b(\gamma) = -0.52\,\gamma + 0.76 \qquad [3.3]$$

$$c(\gamma) = -6{,}050\,\gamma + 13{,}635\ [\mathrm{MPa}^{1}]$$

Figure 3.11 shows how the viscoelastic parameter P_1 evolves as a function of γ. P_2, P_3 and P_4 do not appear to significantly vary (approximately 1–2%), whereas P_1, for γ equal to 0.7, has an increase of approximately 160% relative to the initial value. Thus, unlike softening, the influence of viscoelasticity on the mechanical behavior of the polymer in the short term increases with the degree of oxidation.

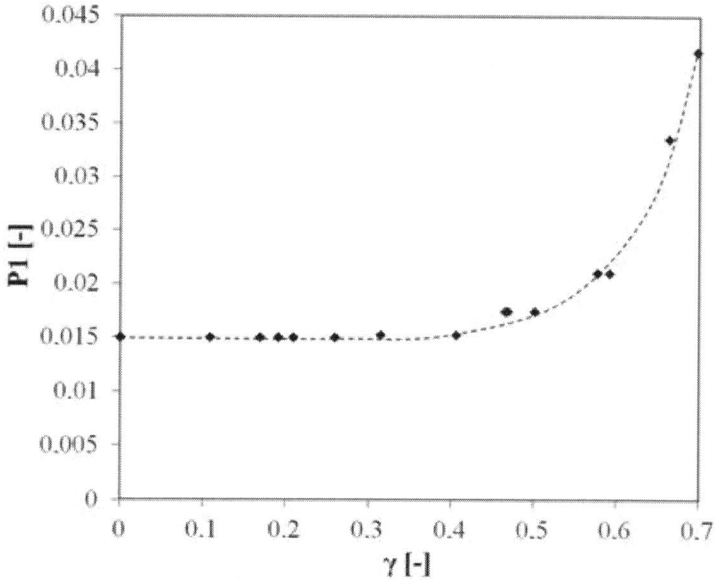

Figure 3.11. *Evolution of the viscoelastic parameter P_1 as a function of γ (oxidized TACTIX resin) (see also [GIG 16a, MIN 13, MIN 14])*

Figure 3.12 shows a comparison between the relaxation spectrum of a virgin sample, in black, and that of a sample oxidized for 7 days under O_2 at 150°C, in red. The variation in P_1 results in a general/normal distribution of the relaxation spectrum, where long relaxation times are also affected.

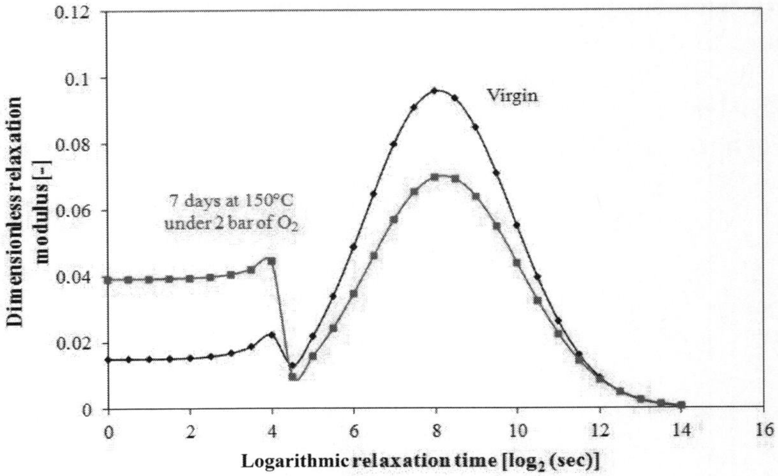

Figure 3.12. *Relaxation spectrum of virgin and oxidized TACTIX polymer (see also [GIG 16a, MIN 13, MIN 14]). For a color version of this figure, see www.iste.co.uk/gigliotti/mechanical.zip*

Finally, in Figure 3.13, the experimental profile measured using IM 10 minutes after the indentation test, on the surface of a sample oxidized for 72 hours at 150°C under 2 bar O_2, is compared with the corresponding numerical profile. It is noted that, once again, the maximum depression is well estimated and the numerical profile is very similar in appearance to that of the experimental profile, which is less regular. This result confirms the validity of the developed constitutive law, which has been shown for the virgin polymer only, but which, according to Figure 3.13, is able to represent the local mechanical behavior of an oxidized sample.

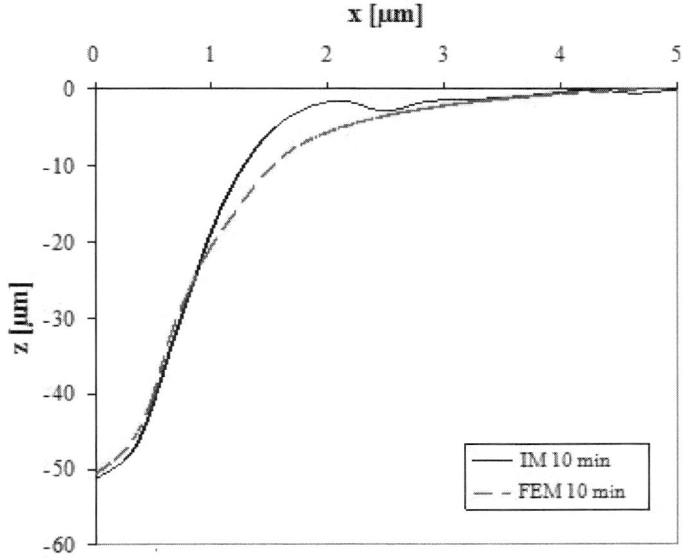

Figure 3.13. *Numerical (FEM) and experimental (IM) profiles of an indentation curve observed 10 minutes after the UMI test on the surface of an oxidized TACTIX sample for 72 hours at 150°C under 2 bar O_2 at a distance of 40 µm from the free edge (see also [GIG 16a, MIN 13, MIN 14]). For a color version of this figure, see www.iste.co.uk/gigliotti/mechanical.zip*

3.2. Identification of inelastic strains of chemical origin by inverse analysis of matrix shrinkage in unidirectional OMCs

For the numerical modeling of unidirectional composites subjected to thermo-oxidative "loading", the conclusions of the experimental study highlight the need to take into account the actual arrangement of the fibers and their interactions. An image representative of the surface area of the UD sample, obtained by IM, was used as a reference (see Figure 3.14(a)). The fibers are drawn by duplicating the image (see Figure 3.14(b)). The surface thus obtained is perforated in correspondence with the fibers (see Figure 3.14(c)), then extruded over a depth of 500 μm (see Figure 3.14(d)). The final three-dimensional geometry was meshed with 262,000 elements of type C3D8R, which are linear elements with reduced integration.

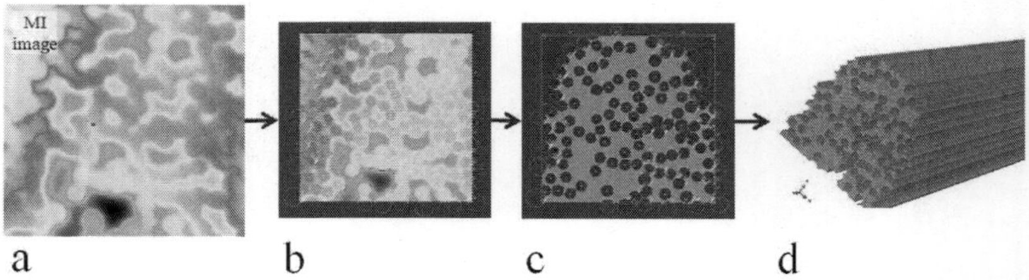

Figure 3.14. *Stages required to obtain the geometric model of the UD composite (see also [GIG 16b, GIG 16c]). For a color version of this figure, see www.iste.co.uk/gigliotti/mechanical.zip*

The mechanical behavior of the various elements of the composite is modeled as follows:

– the fibers are supposed to be perfectly rigid and therefore replaced by a clamped boundary condition. It is also assumed that the coefficient of thermal expansion of the carbon fiber is zero, since it is several orders of magnitude lower than that of the matrix [FIE 02, JIN 07];

– it is assumed that the mechanical behavior of the matrix is locally the same as that of the pure resin, whose law of behavior dependent on ageing was developed in Chapter 2;

– no fiber/matrix interphase is introduced. In the literature, there are many studies [PAL 92, JAY 93, PAP 02, YAN 03], aiming to characterize the zone located near the fibers where an interphase seems to develop, probably due to a cross-linking process of the polymer different from the presence of carbon fibers. Empirical relationships [PAP 02], a kinetic model [YAN 03] and a thermodynamic model based on the calculation of Gibbs free energy [PAL 92] are now available for the modeling of the gradient of interphase properties. On the other hand, these models are derived from purely theoretical considerations and mechanical tests demonstrating their validity (typically spectroscopy tests) are very difficult to implement [SUB 96], since the layer concerned by these phenomena is approximately 100–200 nm in size [ARA 01]. The aim of our study is to globally reproduce the field of matrix shrinkage, mainly related to the mechanical behavior of the matrix; the fiber and the interphase are modeled simultaneously, according to the method proposed in [POC 08].

According to Vu *et al.* [VU 12], the matrix shrinkage observed on the surface of the UD composite is mainly due to the formation of irreversible inelastic strains in the matrix:

– in the initial state, the initial inelastic strain ($\mathbf{E}^{in}$), inducing the matrix shrinkage observed on the surface of the virgin composite, simultaneously takes into account the strain of the matrix by thermal expansion ($\mathbf{E}_T$) and that generated by polishing the surface.

– in the oxidized state, chemical inelastic strains ($\mathbf{E}^{ch}(\gamma)$) develop during oxidation, mainly due to the release of volatile products and water molecules.

These inelastic strains cause a variation in the matrix volume and, therefore, must be introduced into the spherical part of the thermodynamic potential of the polymer matrix, as presented in Chapter 2.

Since the local mechanical properties of the matrix have been identified previously, the only unknowns are the inelastic strains, $\mathrm{tr}\mathbf{E}^{in}$ and $\mathrm{tr}\mathbf{E}^{ch}(\gamma)$, which will subsequently be obtained by minimizing the discrepancies between the matrix shrinkage fields measured by IM on the surface of the sample, and those obtained by numerical simulation.

3.2.1. *Method for identifying inelastic strains in virgin and oxidized OMCs*

To simulate the initial state of the virgin composite, the thermal loading imposed on the numerical model consists of two phases (see Figure 3.15):

– an instantaneous cooling phase, where a temperature change (curing temperature minus ambient temperature) is imposed;

– a long-lasting relaxation phase, where the temperature variation T (and therefore the inelastic strain ε^{in}) remains constant.

This thermal loading corresponds to the experimental procedure which consisted in bringing the composite sample closer to the curing temperature in order to "erase" the previous thermomechanical history, to cool it to ambient temperature and to take observations and measurements after approximately 5 days, or 420,000 seconds.

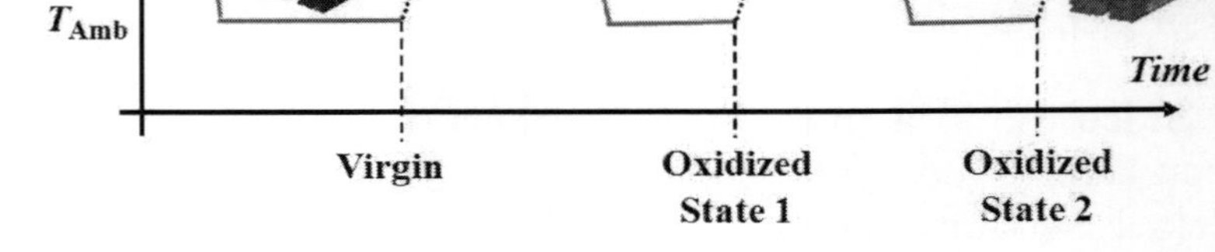

Figure 3.15. *Phases of simulation for the study of matrix shrinkage of the virgin and oxidized material (see also [GIG 16b, GIG 16c]). For a color version of this figure, see www.iste.co.uk/gigliotti/mechanical.zip*

As mentioned in Chapter 1, and based on the results of [VU 12], we assume that ε^{in} is mainly due to the cooling of the material from its curing temperature to ambient temperature. In the numerical simulation, the inelastic strain ε^{in} is therefore proportional to the temperature variation and the proportionality factor is the thermal expansion coefficient of the matrix (α_T), i.e. $\varepsilon^{in} = \alpha_T \, \Delta T$.

For the oxidized state, the structure of the numerical simulation is illustrated in Figure 3.15. A numerical calculation, composed of two phases, is associated with each oxidized state of the experimental study:

– an instantaneous cooling phase of the ageing temperature to ambient temperature;

– a relaxation phase lasting 600 seconds, where the strain ε^{in} remains constant. The duration of the relaxation phase corresponds, once again, to the time required to carry out the experimental observation, after the cooling of the sample.

The mechanical properties of the matrix, as well as their gradient in the oxidized layer, are introduced in the numerical model by means of a PYTHON script as a function of γ, a parameter indicative of ageing. This script evaluates – as the oxidation progresses – the appropriate value of γ to be given to each node of the mesh, as a function of the distance to the surface exposed to the oxidizing environment and the ageing conditions considered.

Then, the constitutive law of the polymer, as well as the chemical inelastic strains, is supplied to ABAQUS as a function of the local value of γ. Using this approach, simulation can reproduce the gradients of properties and strains that develop within the sample from the surface exposed to the environment. In Figure 3.15, for example, the geometry of the matrix modeled after oxidation has a color gradient, where blue corresponds to the condition $\gamma = 0$ and the other colors correspond to higher values of γ.

3.2.2. *Identification of inelastic strains and calculation of stresses in virgin OMCs*

In the virgin state, the mechanical properties of the matrix are homogeneous and uniform. The aim of this section is therefore to estimate the initial inelastic strain, which will serve as a reference for evaluating the inelastic chemical strain induced by thermo-oxidation.

Again, we solve an optimization problem, in which the variable ε^{in} must be identified by minimizing the difference between the displacement fields measured by IM (see Figure 3.16(a)) and numerical analysis (see Figure 3.16(b)). For this purpose, the same optimization protocol, set up for the identification of the law of behavior parameters of the resin, was used. The difference between the numerical and experimental values was evaluated in 15 points, chosen in different zones fairly rich in fibers.

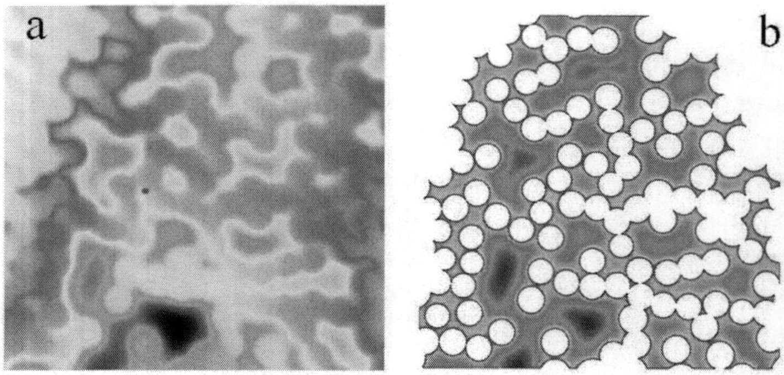

Figure 3.16. *Displacement fields (matrix shrinkage): a) experimental and b) numerical (surface of the virgin UD composite) (see also [GIG 16b, GIG 16c]). For a color version of this figure, see www.iste.co.uk/gigliotti/mechanical.zip*

Using this approach, the value found is $\varepsilon^{in} = -10^2$. In Figure 3.17(a), the four shrinkage profiles (A, B, C and D) corresponding to a fiber-to-fiber distance of 20 μm, measured experimentally, are compared with the numerical profiles; in Figure 3.17(b), estimated maximum matrix shrinkages were compared with measurements for different fiber distances (up to 30 μm). It is interesting to note that with a single value of ε^{in}, the entire shrinkage profile is correctly reproduced.

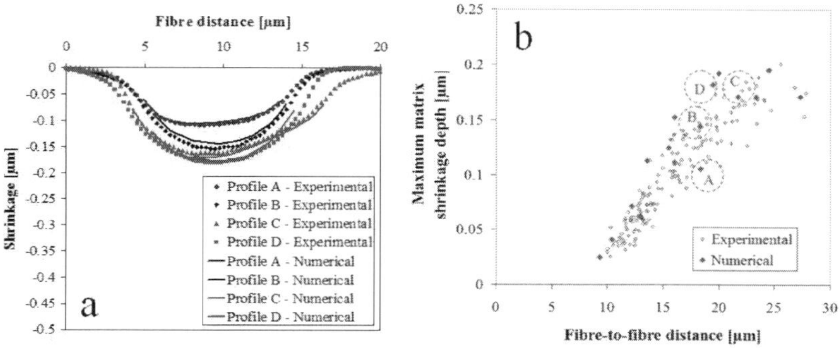

Figure 3.17. *a) Numerical and experimental profiles for a distance between fibers of 20 μm. b) Maximum matrix shrinkage as a function of the distance between fibers (virgin UD composite) (see also [GIG 16b, GIG 16c]). For a color version of this figure, see www.iste.co.uk/gigliotti/mechanical.zip*

This last result confirms that, on the one hand, the modeled zone has an extension that can sufficiently represent the behavior of the composite and, on the other hand, the scatter of the experimental points is not due to the measurement technique but again to the random arrangement of the fibers in the material. Although it is not possible to separate the "thermal" and "polishing" components from the overall inelastic strain ε^{in}, taking the value of $\Delta T = -190°C$ ($T_{curing} - T_{amb}$), a value of $\alpha_T \approx 53 \times 10^{-6}°C^{-1}$ is obtained for the thermal expansion coefficient of the matrix, a typical value for polymers in this family. This suggests that the predominant source of the matrix shrinkage observed at the surface of the virgin composite is in the temperature change from cure to room temperature and that polishing has a secondary effect instead.

The viscous character of the mechanical behavior of the matrix is observed by the relaxation of the stresses and in the evolution of the matrix shrinkage during the simulation time. In Figure 3.18, the evolution of profiles B and D is plotted as a function of simulation time from t = 0 until 420,000 seconds. The matrix shrinkage decreases by approximately 10% in the first 420,000 seconds of simulation and we find that, beyond this, there is little evolution.

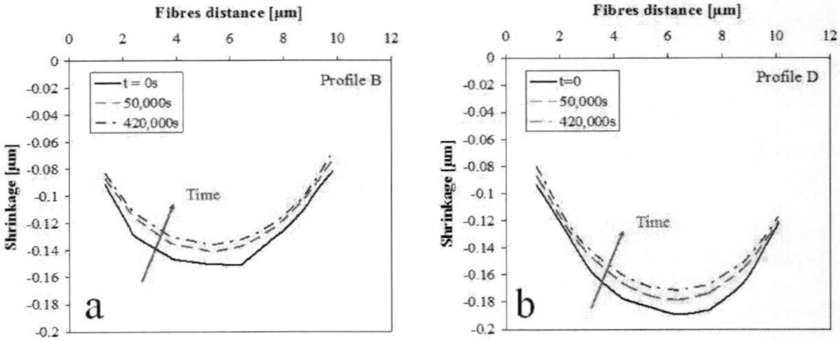

Figure 3.18. *Evolution of numerical matrix shrinkage of profiles a) B and b) D as a function of simulation time (virgin UD composite) (see also [GIG 16b, GIG 16c]). For a color version of this figure, see www.iste.co.uk/gigliotti/mechanical.zip*

The numerical model allows the von Mises stress fields to be obtained immediately after cooling the composite (see Figure 3.19(a)) and after 420,000 seconds (see Figure 3.19(b)), the time required to reach a stabilized state.

From this figure, we can see that:

– there is significant stress relaxation, as we move from an initial maximum value of approximately 130 MPa (see Figure 3.19(a)) to a very low relaxed value of approximately 0.05 MPa (see Figure 3.19(b)). These values are obviously indicative since, in the numerical model, the fibers have not been modeled, but rather replaced by a clamped boundary condition, which prevents calculation of reliable values at the fiber/matrix interfaces: these values depend on the mesh refinement – due to the singularity of the free edge – and increase with it;

– just after cooling (see Figure 3.19(a)), the areas where the stresses are greatest are the matrix-rich areas and, in particular, where the distances between fibers are the greatest.

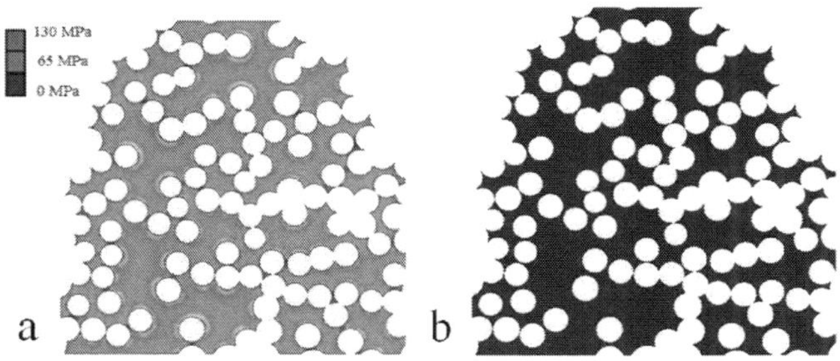

Figure 3.19. *Von Mises stress field in the virgin UD composite: a) just after cooling from the curing temperature and b) after 420,000 seconds of relaxation (see also [GIG 16b, GIG 16c]). For a color version of this figure, see www.iste.co.uk/gigliotti/mechanical.zip*

A unique value of inelastic strain – constant throughout the composite – thus allows the initial field of displacement observed experimentally to be reproduced correctly. Taking into account the actual distribution of the fibers, the exact shapes of the four profiles chosen by way of example can be obtained numerically and, simultaneously, the scatter of the "matrix shrinkage versus fiber distance" curve is obtained. The highest stresses are obtained immediately after the material cooling phase, and then left to relax in time, due to the viscous behavior of the matrix. These stresses are located where the composite is matrix rich and where the distances between fibers are the greatest. At the same time, matrix shrinkage decreases slightly, demonstrating the importance of taking into account a relaxation phase in the numerical simulation before comparing the numerical profiles with the measurements by IM.

3.2.3. *Identification of inelastic strains in oxidized OMCs*

Following oxidation, the mechanical properties of the matrix evolve. At the same time, inelastic strains of chemical origin develop. According to the previous section, the strain ε^{in} has been identified with a value of -10^{-2} and only the value of $tr\mathbf{E}^{ch}(\gamma)$ remains to be identified for the various oxidized states. By introducing a scalar inelastic strain value ($\varepsilon^{ch}(\gamma)$), the expression $tr\mathbf{E}^{ch}(\gamma)$ can be rewritten as follows:

$$tr\mathbf{E}^{ch}(\gamma) = 3\,\varepsilon^{ch}(\gamma) \qquad [3.4]$$

Moreover, $\varepsilon^{Ch}(\gamma)$ and γ are assumed to have the same gradient, as a function of the distance to the surface exposed to the environment (z) and, therefore, for given ageing conditions, we can write:

$$\varepsilon^{ch}(\gamma) = \varepsilon^{ch}_{max}(t)\gamma(z) \qquad [3.5]$$

where $\varepsilon^{ch}_{max}(t)$ is a proportionality coefficient representing inelastic strain at the surface, dependent only on the ageing time, for a given temperature and partial pressure of O_2.

As for the evaluation of ε_{In}, presented in the previous section, the strain $\varepsilon^{ch}_{max}(t)$ can be evaluated using an optimization algorithm.

In Figure 3.20, the shrinkage profiles obtained for configuration D are compared with the measurements for two different ageing durations under 2 bar O_2 at 150°C: after 30 hours with $\varepsilon^{ch}_{max}(t) = -0.6 \times 10^{-2}$ and 42 hours with $\varepsilon^{ch}_{max}(t) = -1.1 \times 10^{-2}$.

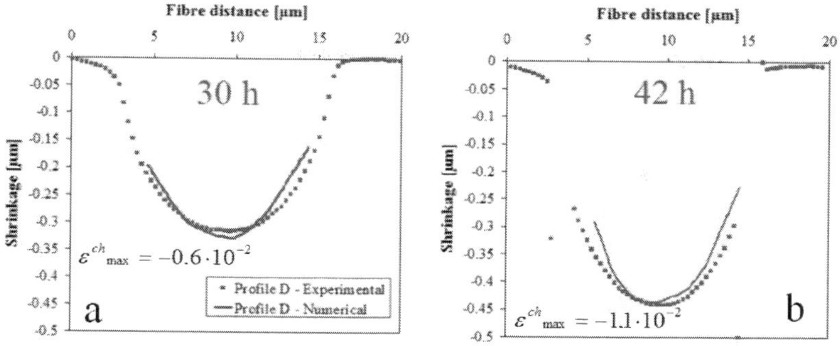

Figure 3.20. *Comparison between the experimental and numerical shrinkage profile "D", after 30 hours and 42 hours of ageing under 2 bar O_2 at 150°C (oxidized UD composite) (see also [GIG 16b, GIG 16c])*

Moreover, using the same values of $\varepsilon^{ch}_{max}(t)$, the complete shrinkage curves are correctly reproduced: in Figure 3.21, the maximum numerical matrix shrinkage was compared with measurements of distances between fibers (up to 40 µm) for two oxidation times (30 hours in Figure 3.21(a) and 42 hours in Figure 3.21(b)), showing that we can also obtain experimental values (hollow symbols) and their scatter.

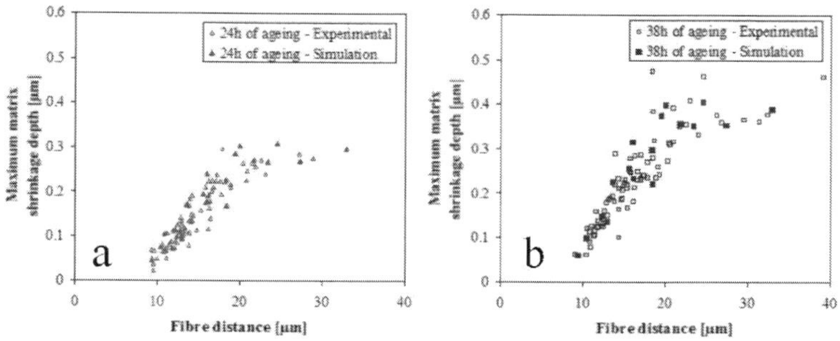

Figure 3.21. *Numerical/experimental comparison of maximum matrix shrinkage as a function of the distance between fibers after: a) 24 hours and b) 38 hours of ageing under 2 bar of O_2 at 150°C (oxidized UD composite) (see also [GIG 16b, GIG 16c]). For a color version of this figure, see www.iste.co. uk/gigliotti/ mechanical.zip*

Evolutions of total and chemical inelastic strains (defined as the sum of initial inelastic strain and surface chemical strain (see equation [3.6])) are presented as a function of γ_{max} in Figure 3.22:

$$\varepsilon^{max}(t) = \varepsilon^{In} + \varepsilon^{ch}_{max}(t) \qquad [3.6]$$

It is observed that the "weight" of the initial strain is important and that the two regimes observed previously can always be distinguished: pre-debonding (for $\gamma_{max} < 0.56$) and post-debonding (for $\gamma_{max} > 0.56$). This value corresponds to the critical value leading to the onset of debonding on 50% of the observed surface. We also identify the critical value of the corresponding total strain $\varepsilon^{max}_{c} = -1.7 \times 10^{-2}$, which does not depend on local fiber configurations or distances between fibers but represents a threshold value beyond which half of the surface is damaged.

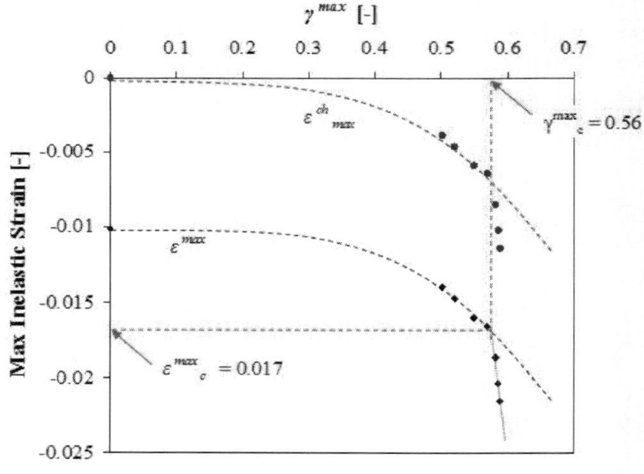

Figure 3.22. *Evolution of chemical (blue) and total inelastic strain (black) during ageing under 2 bar O_2 at 150°C (oxidized UD composite) (see also [GIG 16b, GIG 16c]). For a color version of this figure, see www.iste.co.uk/gigliotti/mechanical.zip*

Note, however, that "post-debonding" inelastic strains are obtained using a numerical model that does not take into account fiber/matrix debonding. As a result of the onset of this damage, the clamped boundary condition between the fiber and the matrix is no longer realistic in the damaged zones and, consequently, this leads to overestimating the total inelastic strain value identified. This aspect will be more broadly discussed in the next section.

In conclusion, it appears that a single value of ε_{max}^{ch}, for a given ageing time, allows us to correctly reproduce the entire shrinkage profile measured by IM. This makes it possible to conclude that the shrinkage profile results solely from the "structure" effect associated with the presence of the fibers, which do not in any way interact with the oxidation phenomena. This observation would allow the use of simplified models for the prediction of shrinkage induced by the thermo-oxidation, models which take into account only this structural effect. Such a simplified predictive model, based on the Rayleigh–Ritz method, will be presented at the end of this chapter.

3.2.4. *Validating the identification of inelastic strains of chemical origin*

In the previous section, chemical strains were identified by an inverse analysis of the matrix shrinkage measured by IM on the surface of UD composite samples for different oxidation times at 150°C under 2 bar O_2. Here, we assumed that the distribution of chemical strains linearly followed that of the phenomenological phenomenon of ageing γ (see equation [3.5]). Figure 3.23 shows the expected gradient of chemical strains ε^{ch} (associated with equation [3.5]) as a function of the distance to the surface exposed to the environment for 24 hours, 30 hours, 36 hours and 42 hours of thermo-oxidative ageing.

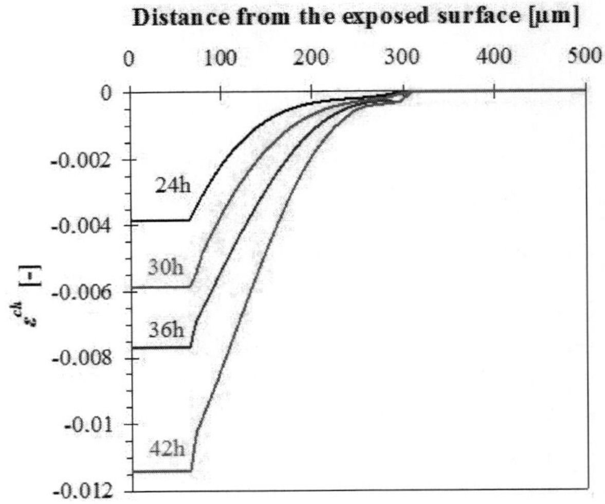

Figure 3.23. *Strain profiles of chemical origin (equation [3.5]) as a function of the distance to the surface exposed to the environment for 24 hours (black), 30 hours (red), 36 hours (blue) and 42 hours (green) of ageing under 2 bar O_2 at 150°C (see also [GIG 16b, GIG 16c]). For a color version of this figure, see www.iste.co.uk/gigliotti/mechanical.zip*

To validate this hypothesis and the expected shape of the chemical strain gradient, a numerical model of a pure resin sample was set up (see Figure 3.24) in order to calculate, with this hypothesis, the surface shrinkage induced by thermo-oxidation and compare it with that measured by IM.

As before, the mechanical properties of the resin and the inelastic strains are related to the parameter γ and are compared with the shrinkage measured on the surface of a sample oxidized for 24 hours under 2 bar of oxygen at 150°C. The gradient of γ, corresponding to this oxidized state, was introduced into the simulation using a FORTRAN script. Figure 3.24 shows the geometry of the sample (dimensions 20 × 5 × 0.5 mm), symmetry conditions imposed, surface exposed to the environment and surface area observed for shrinkage.

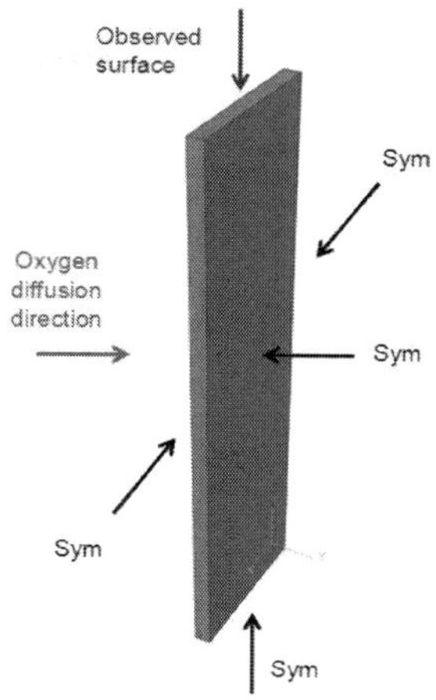

Figure 3.24. *Numerical model for the simulation of surface shrinkage of a pure oxidized resin sample (see also [GIG 16b, GIG 16c]). For a color version of this figure, see www.iste.co.uk/gigliotti/mechanical.zip*

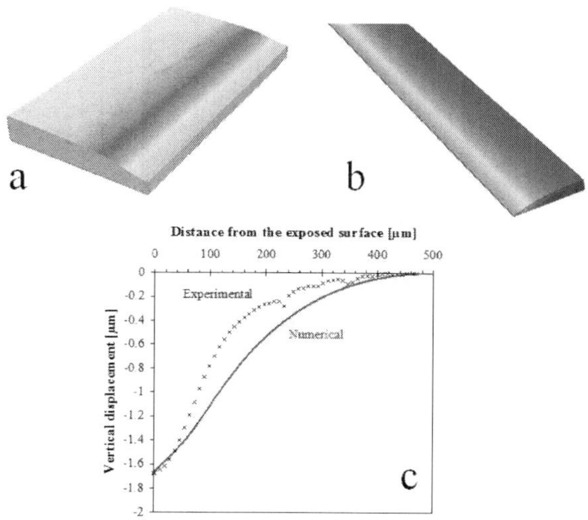

Figure 3.25. *a) IM image of the surface of a TACTIX resin sample oxidized for 24 hours at 150°C under 2 bar O_2; b) numerical simulation result; c) comparison of experimental and numerical shrinkage profiles (see also [GIG 16b, GIG 16c]). For a color version of this figure, see www.iste.co.uk/gigliotti/mechanical.zip*

The simulation result is shown in Figure 3.25(b) against the experimental measurements by IM (see Figure 3.25(a)). Both profiles are qualitatively similar. To obtain a quantitative estimate, the surface profiles (numerical and experimental) are plotted in Figure 3.25(c).

The "numerical" shrinkage is similar to the experimental values: in both cases, the shrinkage involves a layer of approximately 400 μm, with a maximum value of approximately 1.7 μm. However, there is a slight difference in the shape of the shrinkage profile, the numerical model overestimating the vertical displacement of the surface at the center of the oxidized layer. Despite this difference, the "overall" shrinkage is reasonably well reproduced and this study allows us to validate the inelastic properties and strain gradients identified in the previous sections.

3.2.5. *Numerical simulation of stresses induced by thermo-oxidation in UD OMCs*

The numerical model provides access, by calculation, to the distribution of the stresses associated with chemical strains, as well as their evolution over time.

In order to interpret the damage phenomena occurring at this scale (fiber/matrix debonding), stress fields are analyzed at different ageing times. For this study, the mappings from the numerical simulations are compared with experimental observations. In particular, the superposition of the von Mises stress numeric field and the shrinkage field, measured by IM for a configuration leading to the appearance of debonding in certain places (and at certain interfaces), allows us to identify a constraint "threshold", which can be considered as a critical damage onset stress. However, this value is merely indicative because its estimate at the fiber/matrix interface depends on the mesh density.

To begin, we consider a sample aged for 38 hours under 2 bar O_2 at 150°C, with a γ_{max} value close to 0.56 (critical value) and several damaged sites (fiber/matrix debonding), particularly where the fiber volume fraction is low. For the oxidation level considered, the shrinkage field observed by IM is shown in Figure 3.26(a).

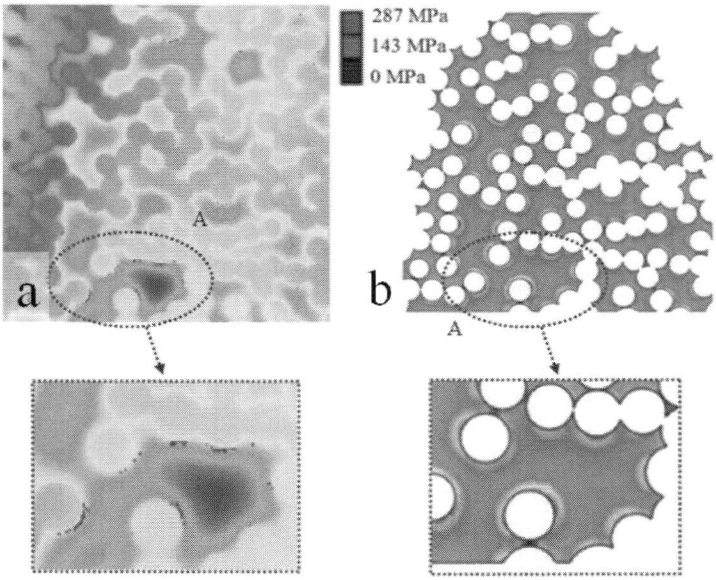

Figure 3.26. *a) Image obtained by IM of the UD composite surface oxidized for 38 hours under 2 bar O_2 at 150°C – the points not measured correspond to debonding; b) corresponding stress field, just after cooling the composite (oxidized UD composite) (see also [GIG 16b, GIG 16c]). For a color version of this figure, see www.iste.co.uk/gigliotti/mechanical.zip*

In this image, zone A is the part of the surface observed experimentally to be the most matrix rich and the one where most of the unmeasured points are located (violet points in Figure 3.26(a)). Figure 3.26(b) illustrates the corresponding stress field, calculated just at the end of the cooling phase, the most critical moment before viscoelastic stress relaxation. It is noted that, in zone A, the stresses are high, in particular for the distances between the largest fibers. By comparing the stress values with the presence of unmeasured points on the surface of the sample (see Figure 3.26(a)), a "threshold" value of 240 MPa is identified. Moreover, these calculations show that, after 38 hours of ageing at 150°C under 2 bar O_2, the mean value of the von Mises stress, estimated on the surface considered and immediately after cooling, is approximately 115% of that obtained on the surface of the virgin material.

The most oxidized sample of this study is now considered, that is to say aged for 42 hours under 2 bar of O_2 at 150°C. This sample has a γ_{max} value slightly greater than 0.56, and fiber/matrix debonding is observed over almost the entire surface by IM (see Figure 3.27(a)). In this case, the mean value of the von Mises stress increased by approximately 207% with respect to the virgin material. By applying the previously identified threshold ($\sigma^{VM}{}_s$ = 240 MPa) to the corresponding von Mises stress field, the "critical" areas predicted by the numerical model (red in Figure 3.27(b)) are almost the same as those in which damage is observed experimentally (see Figure 3.27(a)). This result is very interesting, since it confirms that fiber/matrix debonding is related to a "structural effect".

Now, two questions arise: can the sudden variation in the total inelastic strain ε^{max}, when γ reaches the critical value $\gamma^{max}{}_c = 0.56$ (see Figure 3.22), be explained by the development of fiber/matrix debonding? And to answer this question, how deep is the debonding in the sample from the surface exposed to the environment (value not experimentally measurable)? To answer these questions, a dedicated study should be conducted. This study is still one of the outlooks of this research.

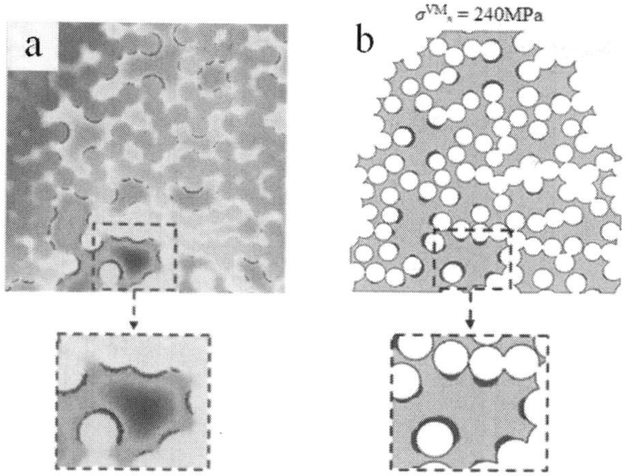

Figure 3.27. *a) Image obtained by IM on the surface of an oxidized UD composite 42 hours under 2 bar O_2 at 150°C; b) zones where the corresponding von Mises stress field goes beyond the 240 MPa threshold (see also [GIG 16b, GIG 16c]). For a color version of this figure, see www.iste.co.uk/gigliotti/mechanical.zip*

3.2.6. *Rayleigh–Ritz approach for approximate computation of matrix shrinkage in OMCs*

The measurements by IM show that the arrangement of the fibers on the surface of a composite is not uniform and that the matrix shrinkage strongly depends on the local configuration and the distance between fibers.

However, according to [VU 12], if the distribution of the fibers near the matrix-rich regions is more meticulously observed, it can be seen that the matrix shrinkage can be considered two-dimensional, making it possible to model these zones with a 2D displacement field (see Figure 3.28). This observation opens the way for the implementation of an approximate model, based on the Rayleigh–Ritz method, for the simulation of shrinkage fields.

In the Rayleigh–Ritz method, the matrix shrinkage field – induced by thermo-oxidation – is calculated by minimizing the total potential energy, written in the purely elastic case:

$$\mathcal{E} = \int_V (\frac{1}{2} (\mathbf{E}: \mathbb{C}\, \mathbf{E} - \mathbf{E}^{TH}: \mathbb{C}\, \mathbf{E} - \mathbf{E}^{CH}: \mathbb{C}\, \mathbf{E}))dV \quad [3.7]$$

where $\mathbb{C}$ is the elasticity tensor, $\mathbf{E}$ the mechanical strain tensor, $\mathbf{E}^{TH}$ the inelastic strain tensor of thermal origin and $\mathbf{E}^{CH}$ the inelastic strain tensor of chemical origin. More precisely, in the Rayleigh–Ritz method, the following conditions are valid at equilibrium:

$$\delta\mathcal{E} = 0 \text{ for equilibrium} \quad [3.8]$$

$$\delta^2\mathcal{E} > 0 \text{ for stable equilibrium}$$

In the Rayleigh–Ritz approach, the displacement fields are discretized using an approximation of the following type:

$$\mathbf{u}^{RR}(x, y, z) = \sum_{i=1}^{q} \mathbf{a}_i \chi_i(x, y, z) \quad [3.9]$$

where $\chi_i(x, y, z)$ represent functions chosen *a priori*, which must satisfy the kinematic-type boundary conditions (kinematically admissible fields). Polynomial functions, trigonometric functions and Chebyshev polynomials are generally appropriate functions. The total potential energy thus takes the approximate form:

$$\mathcal{E}^{RR}(\mathbf{u}^{RR}(x, y, z)) = \mathcal{E}^{RR}(\mathbf{a}_i) \quad [3.10]$$

In addition, the conditions of equilibrium and stability become:

$$\frac{\partial \mathcal{E}^{RR}(\mathbf{a}_i)}{\partial \mathbf{a}_i} = 0, i = 1 \dots q \quad [3.11]$$

$$\frac{\partial^2 \mathcal{E}^{RR}(\mathbf{a}_i)}{\partial \mathbf{a}_i \partial \mathbf{a}_i} > 0$$

To simulate the matrix shrinkage between fibers induced by the thermal expansion alone in UD composites, the following displacement field can be used (with respect to a reference system, such as that illustrated schematically in Figure 3.28):

$$\mathbf{u}^{\mathrm{RR}}(\mathrm{x},\mathrm{y},\mathrm{z}) \equiv \begin{pmatrix} 0 \\ 0 \\ Ax(x-L)\frac{(H-z)}{H} \end{pmatrix} \qquad [3.12]$$

where L is the distance between fibers and A a constant to be determined by imposing the equilibrium and stability conditions (see equation [3.11]).

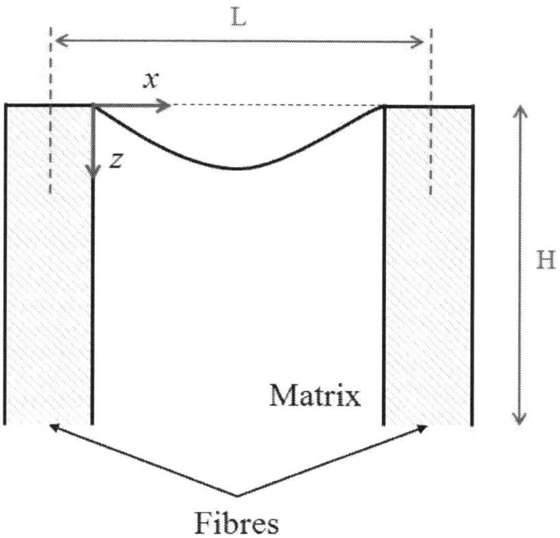

Figure 3.28. *Schematic configuration of the Rayleigh–Ritz simulation of matrix shrinkage close to the surface exposed to the environment (see also [GIG 16b, GIG 16c]). For a color version of this figure, see www.iste.co.uk/gigliotti/mechanical.zip*

In Figure 3.28, H represents the thickness of a layer in the sample where vertical displacement occurs. Indeed, according to equation [3.12], it is assumed that the out-of-plane displacement evolves linearly with z: on the surface exposed to the environment, the displacement is maximum, whereas at a certain depth – expressed precisely by the parameter H – it is zero. This hypothesis was validated by numerical simulations with finite elements, which allowed us to associate with H a value between 10 μm and 15 μm, according to the ageing conditions and the distance between fibers.

Applying these conditions gives the following results:

$$\frac{\partial \varepsilon^{RR}(A)}{\partial A} = 0 \Rightarrow A = \frac{30H(1+\nu^R)(\varepsilon^{in})}{5H^2(2\nu^R-1)+6L^2(\nu^R-1)} \quad [3.13]$$

$$\frac{\partial^2 \varepsilon^{RR}(A)}{\partial A^2} > 0 \Rightarrow \frac{E^R L^3}{180}[(5H^2 + 6L^2) - \nu^R(10H^2 + 6L^2)] \frac{1}{H(1-2\nu^R)(1+\nu^R)} > 0 \Rightarrow 0 < \nu^R < 0.5$$

where ν^R and E^R represent the relaxed values of Poisson's coefficient and Young's modulus of the material and ε^{in} is the initial inelastic strain. Equation [3.13b] has been *a priori* satisfied for the classical values of Poisson's coefficient of polymers.

Finally, in the initial state, the out-of-plane displacement, w (x, y, z), depends only on the initial inelastic strain, ε^{In}, of the distance between fibers L, of the geometric parameter H and the relaxed Poisson's coefficient. The results of the numerical calculations obtained by taking into account three different initial inelastic strain values (0.5×10^{-2}, 0.95×10^{-2} and 1.5×10^{-2}) are illustrated in Figure 3.29 and compared with their corresponding experimental points, for a single value of H (H = 10 μm) and ν^R ($\nu^R = 0.3$).

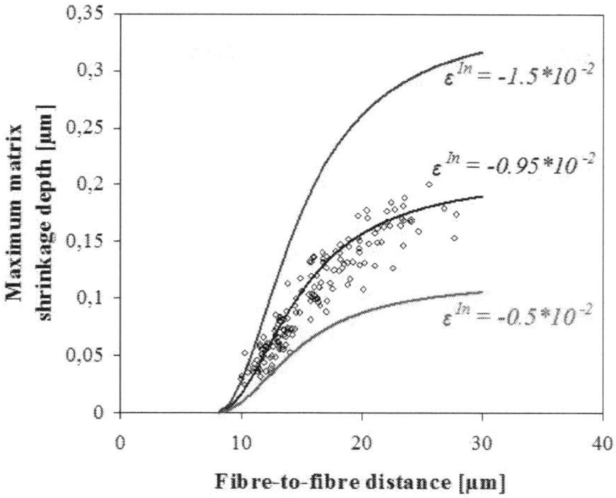

Figure 3.29. *Comparison of the maximum matrix shrinkage as a function of distance between fibers observed experimentally on the surface of a virgin UD composite (points) with that simulated using the Rayleigh–Ritz model for three different initial inelastic strain values ε^{in} (continuous lines) (see also [GIG 16b, GIG 16c]). For a color version of this figure, see www.iste.co.uk/gigliotti/mechanical.zip*

For value $\varepsilon^{in} = -0.95 \times 10^{-2}$, the numerical curve is reasonably in agreement with the experimental points. This value is extremely close to that identified for initial inelastic strains using the FE model (-1×10^{2}), with a deviation of 5%. On the other hand, this model does not provide any information concerning the scatter of points.

In the presence of chemical strains, parameter A becomes:

$$A = \frac{30H(1+\nu^R)(\varepsilon^{in}+\varepsilon^{ch}_{max}(t))}{5H^2(2\nu^R-1)+6L^2(\nu^R-1)} \qquad [3.14]$$

where $\varepsilon^{ch}_{max}(t)$ represents the inelastic strain of maximum chemical origin and depends on the ageing time t.

In Figure 3.30, experimental matrix shrinkage is plotted as a function of the distance between fibers, measured in the initial state and after 30 hours and 42 hours of ageing at 150°C under 2 bar O_2, and compared with the results of the Rayleigh–Ritz model obtained for three different values of $\varepsilon_{max}^{ch}(t)$:

– for the composite in the virgin state (black line), the only component of inelastic strain retained is the strain ε^{in}, i.e. $\varepsilon_{max}^{ch}(t) = 0$;

– the matrix shrinkage of the composite aged for 30 hours (red line) was reproduced numerically, considering a strain $\varepsilon_{max}^{ch}(t)$ of approximately -0.55×10^2;

– the matrix shrinkage of the composite aged for 42 hours (blue line) was reproduced with a strain $\varepsilon_{max}^{ch}(t)$ of approximately -1×10^2.

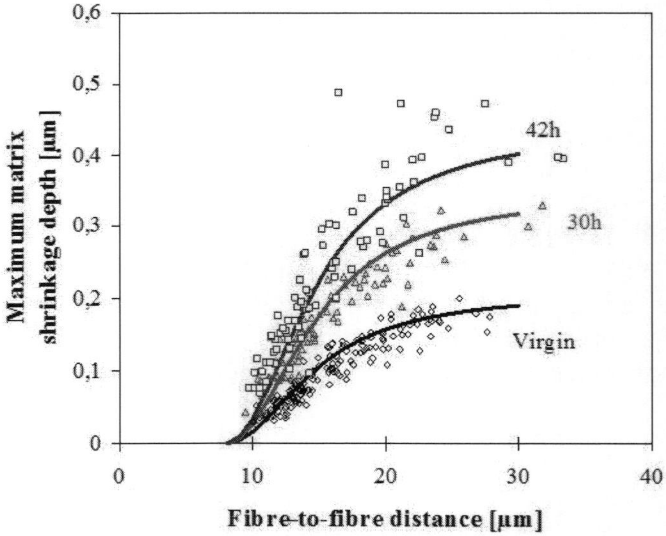

Figure 3.30. *Comparison of the maximum matrix shrinkage as a function of the fiber-to-fiber distance observed experimentally on the surface of a virgin UD composite and oxidized for 30 hours and 42 hours at 150°C under 2 bar O_2 (points), with that simulated using the Rayleigh–Ritz model (continuous lines) (see also [GIG 16b, GIG 16c]). For a color version of this figure, see www.iste.co.uk/gigliotti/mechanical.zip*

In Figure 3.30, the curves from the model are similar to the corresponding experimental points. The total inelastic strains ($\varepsilon_{max}^{ch}(t)$) – which are defined as the sum of ε^{in} plus $\varepsilon_{max}^{ch}(t)$ – identified by the Rayleigh–Ritz method are compared in Table 3.2 with the values computed using the FE model as a function of oxidation time. This comparison revealed small discrepancies between the values of the total inelastic strains obtained by the FE model and by the Rayleigh–Ritz method.

Oxidation duration	**Rayleigh–Ritz** – ε_{RR}^{max}	**Finite elements** – ε_{EF}^{max}	**Deviation**
Virgin	0.95×10^{-2}	1.00×10^{-2}	–5%
30 hours	1.50×10^{-2}	1.60×10^{-2}	–6%
42 hours	1.95×10^{-2}	1.89×10^{-2}	3%

Table 3.2. *Total inelastic strains identified by the Rayleigh–Ritz method and by the FE model (see also [GIG 16b, GIG 16c])*

The Rayleigh–Ritz method, although an approximate method, has many advantages over the finite element method. Indeed, the implementation of FE models and their use require relatively long computing times, which are often not compatible with the paces required by manufacturers. The Rayleigh–Ritz method, which provides analytical formulas, allows for the rapid calculation of matrix shrinkage and stress induced by thermo-oxidation in UD composites, and could be used for the optimization and design of components subject to thermo-oxidative ageing.

Conclusion and Perspectives

This section summarizes the main results obtained in this study and the major advances made, as well as highlighting research perspectives.

Effect of thermo-oxidation on the mechanical behavior of polymers and polymer matrix composites

The mechanical behavior of the polymer studied – an epoxy resin of the TACTIX family – was initially characterized by ultra-micro-indentation (UMI) and dynamic mechanical analysis (DMA). The material is very sensitive to thermo-oxidation which induces large variations in the indentation elastic modulus (EIT) and creates thick oxidized layers, even for relatively short ageing durations. Moreover, the observation of an increase in the DMA elastic modulus (E') measured at ambient temperature is associated with the "antiplasticization" phenomenon, typical for this type of polymer. The development of inelastic strains of chemical origin, mainly due to volatile products, leads to strains and stresses in composites. The experimental study of these strains was carried out by using interferometric microscopy (IM) to measure the matrix shrinkage on the surface of thick unidirectional composite (UD) samples. A statistical study correlates the matrix shrinkage with local fiber volume fraction values, revealing a “structural” effect due to the presence of fibers in the vicinity of the matrix; it is the matrix-rich zones which undergo the greatest matrix shrinkage.

Introduction of a new ageing parameter

It has been shown that the local oxidation state of the aged polymer under isothermal conditions (T) depends on the partial pressure of O_2 (p), the ageing time (t) and the position in the oxidized layer (z). To account for this aged condition, a new phenomenological phenomenon of ageing – γ (T, p, t, z) – was defined. It is obtained from the local value of the indentation elastic modulus (EIT), measured experimentally in the unloading phase of the indentation curves, and using the Oliver and Pharr method. According to the literature, given the correlation between the EIT and the concentration of oxidation products, the parameter γ can be considered as a “tracer” of the oxidation chemistry. This parameter was found to be relevant for interpreting experimental observations on the resin and UD composite, including:

– in the oxidized layer of an aged polymer sample, whatever the values of p, t and z, the same indentation curve always corresponds to the same value of γ, suggesting a direct link between the local mechanical behavior of the polymer and the ageing parameter γ;

– in oxidized UD composites, the evolution of matrix shrinkage follows two different regimes, pre/post fiber-matrix debonding. The transition from one regime to the next is abrupt and characterized by a critical value of γ_{max} –

value of γ measured on the surface exposed to the environment – which appears independent of the ageing conditions.

An isothermal phenomenological evolution law has been proposed for γ, which allows us to reproduce numerically the oxidation profiles developing during ageing at 150°C in air at atmospheric pressure (up to 400 hours) and under 2 bar of oxygen (up to 72 hours).

Development and identification of a local behavior law of the polymer at the microscopic scale, taking into account the effects of thermo-oxidation

The local behavior of the virgin TACTIX resin was initially characterized using a new numerical/experimental methodology based on the inverse analysis of two local tests: UMI test and the following of indentation prints over time by IM. Both tests were analyzed with the aim of finite element numerical models using the commercial software ABAQUS. Mechanical behavior at the local scale has been reasonably well-reproduced using an isothermal, robust and powerful nonlinear viscoelastic behavior law that allows the numerical reproduction of a wide variety of tests. The same numerical/experimental approach was then used to characterize the evolution of the behavior law parameters (elastic properties and parameters of softening and viscoelastic behavior) during thermo-oxidative ageing, and to correlate this evolution with the phenomenological ageing parameter, γ. The thermo-oxidation thus leads to a "more rigid" (consistent with the DMA result), less softening material, characterized by a relaxation spectrum that redistributes favoring the relaxation phenomena in the short term.

Identification of thermo-oxidation inelastic strains through the inverse analysis of matrix shrinkage on the surface of UD composites

In order to assess the inelastic strains associated with thermo-oxidation, a new numerical/experimental methodology was developed, based on the inverse analysis of matrix shrinkage on the surface of UD composites, using a finite element numerical model employing the ageing-dependent behavior law of the polymer. The sample actual fiber distribution and volume fraction were found to be a fundamental parameter accounting for the "structural" effect affecting thermo-oxidation-induced matrix shrinkage. Initial inelastic strains, obtained from the shrinkage field of a virgin UD composite, were found to be uniform in the sample. On the other hand, the chemical inelastic strains induced by thermo-oxidation present a gradient in the depth of the sample, which depends on the oxidation conditions. In the numerical simulation, these strains are assumed to be linear functions of the parameter γ. This hypothesis was validated by comparison with the shrinkages measured by IM in oxidized polymer samples.

The implementation of the finite element numerical model and its use require relatively long computing times, which are often not compatible with industrial requirements. This is why another simpler numerical model based on the Rayleigh–Ritz method has been proposed for the calculation of matrix shrinkage induced by thermo-oxidation in UD composites – a model which could be used for the optimization and design of structural parts.

FE numerical simulation of stress fields on the surface of UD composites, and their evolution with thermo-oxidation

The FE model provides access to the distribution of the stresses associated with the inelastic strains on the surface of the UD composite, as well as their evolution during oxidation. The viscous character of the mechanical behavior of the matrix results in a quasi-complete relaxation of the stresses, and the highest stress values occur immediately after the cooling of the material from cure/use to room temperature. Therefore, the most critical time for the onset of fiber/matrix debonding is that at which the sample is removed from the ageing chamber and cooled to room temperature. Moreover, the value of the average von Mises stress increases with ageing time, due to the increase in chemical strains and due to the change in local mechanical properties of the polymer matrix. A detailed analysis of the von Mises stress fields, calculated using the finite element model for the different oxidized states, allows us to discover that fiber/matrix debonding is likely to occur near matrix-rich zones, consistent with the experimental observation.

Using an increase in oxygen partial pressure as an acceleration parameter

An in-depth analysis of the evolution of the surface value of γ (γ_{max}) in air at atmospheric pressure and under oxygen pressure allows us to establish a time/pressure equivalence. To achieve the same level of oxidation, the ageing time at an average oxygen partial pressure is λ-times shorter than that in air at atmospheric pressure. In this study, which involved the TACTIX resin and the associated composite, the phenomenon of oxidation at 150°C under 2 bar O_2 was found to be approximately 7 times quicker than that in air ($\lambda \approx 7$). At the same time, on the surfaces of unidirectional composites, it was observed that the evolution of matrix shrinkage on the surface and the onset of fiber/matrix debonding depended solely on the ageing kinetics of the material surface. The same surface oxidation rate (and therefore the same value of γ on the surface, γ_{max}) is related to similar matrix shrinkage and "damaged" surfaces. Therefore, a "moderate" oxygen partial pressure can be used effectively as a parameter to accelerate thermo-oxidation both when characterizing the evolution of the mechanical properties of the polymer matrix and studying the fiber/matrix debonding on the surface of a UD composite.

Perspectives

The conclusions of this work indicate four main areas for future development:

– A first interesting perspective would be to limit the possible acceleration range and the values of the associated acceleration factor λ depending on the ageing conditions (temperature and partial pressure of O_2). Tests on pure resin samples would therefore be necessary, based on ultra-micro-indentation measurements of the γ profiles of oxidized samples at different temperatures (not too close to the glass transition temperature, so as not to activate other degradation mechanisms), and under different partial pressures of oxygen (while remaining within the "moderate" pressure range). For a given polymer, this information, associated with the oxidation kinetics identified by chemists, should undoubtedly allow us to have a better understanding of the proposed accelerated tests.

– When considering the use of composites at high temperature, a more thorough study of the thermomechanical properties of the polymer matrix is essential. At "hot" temperatures, the variation in mechanical properties can be significant, especially that of the viscous component of the behavior - all the more so as it is at these temperature levels that the degradation phenomena by thermo-oxidation take place. The development of indentation tests at high temperatures and the use of the same numerical/experimental methodology applied in this work would allow for the explanation of the evolution of the constitutive law parameters with temperature and level of oxidation (γ).

– Predicting the damage initiation and propagation conditions, whether by FEM or by other approximate methods, requires a proper knowledge of the critical stress or energy release rate, which vary with the oxidation level. Indeed, some authors have shown that thermo-oxidation leads to embrittlement of the polymers and macromolecular chain scission at the fiber/matrix interfaces. This characterization has not been carried out within the context of this book and it is a very interesting perspective.

– Matrix oxidation/cracking interaction in polymer matrix composites represents an important aspect of mechanical/environmental coupling. Damage developing either during thermo-oxidation or as a result of mechanical stress provides new surfaces directly exposed to the thermo-oxidizing environment, which could self-accelerate the degradation of the material by oxidation. In order to study these interactions in greater detail, both experimental and numerical methodologies need to be developed.

Integrating the models developed for the resin and for the matrix in numerical calculations of complex composite materials/structures (2D/3D textiles, structural parts, etc.) is foreseen, in order to understand their behavior and their optimization for their use in realistic environmental conditions.

Bibliography

[ARA 01] ARAYASANTIPARB D., MCKNIGHT S., LIBERA M., "Compositional variation within the epoxy/adherend interphase", *Journal of Adhesion Science and Technology*, vol. 15, pp. 1463–1484, 2001.

[BAT 79] BATAILLE J., KESTIN J., "Irreversible processes and physical interpretation of rational thermodynamics", *Journal of Non-Equilibrium Thermodynamics*, vol. 4, pp. 229–258, 1979.

[BRE 09] VAN BREEMEN L., ENGELS T., PELLETIER C. *et al.*, "Numerical simulation of flat-tip micro-indentation of glassy polymers: Influence of loading speed and thermodynamic state", *Philosophical Magazine,* vol. 89, pp. 677–696, 2009.

[CIN 02] CINQUIN J., "Les composites en aérospatiale", *Techniques de l'Ingénieur*, 2002.

[COL 05] COLIN X., MAVEL A., MARAIS C. *et al.*, "Interaction between cracking and oxidation in organic matrix composites", *Journal of Composite Materials*, vol. 39, pp. 1371–1389, 2005.

[COL 12] COLIN X., VERDU J., "Mechanisms and kinetics of organic matrix thermal oxidation", in POCHIRAJU K.V., TANDON G.P., SCHOEPPNER G.A. (eds), *Long-Term Durability of Polymer Matrix Composites,* Springer, New York, 2012.

[CUN 01] CUNAT C., "The DNLR approach and relaxation phenomena. Part 1: Historical account and DNLR formalism", *Mechanics of Time-Dependent Materials*, vol. 5, pp. 39–65, 2001.

[DAS 09] DASSAULT SYSTEMES SIMULIA CORP., Abaqus, version 6.9., United States, 2009.

[DEC 03] DECELLE J., HUET N., BELLENGER V., "Oxidation induced shrinkage for thermally aged epoxy networks", *Polymer Degradation and Stability*, vol. 81, pp. 239–248, 2003.

[DEG 62] DE GROOT A., MAZUR P., *Non-Equilibrium Thermodynamics*, North-Holland, Amsterdam and Wiley, New York, 1962.

[DES 01] DE SIMONE A., MARIGO J.J., TERESI L., "A damage mechanics approach to stress softening and its application to rubber", *European Journal of Mechanics – A/Solids*, vol. 20, pp. 873–892, 2001.

[ECK 40] ECKART C., "The thermodynamics of irreversible processes, II. Fluid mixtures", *Physical Review*, vol. 58, pp. 269–275, 1940.

[FAY 05] FAYOLLE B., VERDU J., "Vieillissement physique des matériaux polymères", *Techniques de l'Ingénieur*, 2005.

[FIE 02] FIEDLER B., HOJO M., OCHIAI S., "The influence of thermal residual stresses on the transverse strength of CFRP using FEM", *Composites Part A: Applied Science and Manufacturing*, vol. 33, pp. 1323–1326, 2002.

[GIG 16a] GIGLIOTTI M., MINERVINO M., LAFARIE-FRENOT M.C. *et al.*, "Effect of thermo-oxidation on the local mechanical behaviour of epoxy polymer materials for high temperature applications", *Mechanics of Materials*, vol. 101, pp. 118–135, 2016.

[GIG 16b] GIGLIOTTI M., MINERVINO M., LAFARIE-FRENOT M.C., "Thermo-oxidative induced shrinkage in organic matrix composites for high temperature applications: effect of fibre arrangement and oxygen pressure", *Composite Structures*, vol. 146, pp. 176–186, 2016.

[GIG 16c] GIGLIOTTI M., MINERVINO M., LAFARIE-FRENOT M.C., "Assessment of thermo-oxidative induced chemical strain by inverse analysis of shrinkage profiles in unidirectional composites", *Composite Structures*, vol. 157, pp. 320–336, 2016.

[GLA 71] GLANSDORFF P., PRIGOGINE I., *Thermodynamic Theory of Structure, Stability and Fluctuations*, Wiley-Interscience, New York, 1971.

[GRA 15] GRANDIDIER J.C., OLIVIER L., LAFARIE-FRENOT M.C. *et al.*, "Modeling the pressure dependent solubility in a thermoset resin for simulating pressure accelerated thermo-oxidation tests", *Mechanics of Materials*, vol. 84, pp. 44–54, 2015.

[GUR 10] GURTIN M.E., FRIED E., ANAND L., *The Mechanics and Thermodynamics of Continua*, Cambridge University Press, New York, 2010.

[JAY 93a] JAYARAMAN K., REIFSNIDER K.L., SWAIN R.E., "Elastic and thermal effects in the interphase. 1 – Comments on characterization methods", *Journal of Composites Technology & Research*, vol. 15, pp. 3–13, 1993.

[JAY 93b] JAYARAMAN K., REIFSNIDER K.L., SWAIN R.E., "Elastic and thermal effects in the interphase. 2 – Comments on modeling studies", *Journal of Composites Technology & Research*, vol. 15, pp. 14–22, 1993.

[JIN 07] JIN K.K., OH J.H., HA S.K., "Effect of fiber arrangement on residual thermal stress distributions in a unidirectional composite", *Journal of Composite Materials*, vol. 41, pp. 591–611, 2007.

[JOH 13] JOHLITZ M., LION A., "Chemo-thermomechanical ageing of elastomers based on multiphase continuum mechanics", *Continuum Mechanics and Thermodynamics*, vol. 25, pp. 605–624, 2013.

[JOH 14] JOHLITZ M., DIERCKS N., LION A., "Thermo-oxidative ageing of elastomers: a modelling approach based on a finite strain theory", *International Journal of Plasticity*, vol. 63, pp. 138–151, 2014.

[KER 08] KERMOUCHE G., LOUBET J.L., BERGHEAU J.M., "Extraction of stress-strain curves of elastic-viscoplastic solids using conical/pyramidal indentation testing with application to polymers", *Mechanics of Materials*, vol. 40, pp. 271–283, 2008.

[LAF 04] LAFARIE-FRENOT M.C., ROUQUIE S., "Influence of oxidative environments on damage in c/epoxy laminates subjected to thermal cycling", *Composites Science and Technology*, vol. 64, pp. 1725–1735, 2004.

[MAU 99] MAUGIN G.A., *The Thermodynamics of Nonlinear Irreversible Behaviors – An Introduction*, World Scientific Publishing, Singapore, 1999.

[MIN 13] MINERVINO M., GIGLIOTTI M., LAFARIE-FRENOT M.C. *et al.*, "The effect of thermo-oxidation on the mechanical behaviour of polymer epoxy materials", *Polymer Testing*, vol. 32, pp. 1020–1028, 2013.

[MIN 14] MINERVINO M., GIGLIOTTI M., LAFARIE-FRENOT M.C. *et al.*, "A coupled experimental/numerical approach for the modelling of the local mechanical behaviour of epoxy polymer materials", *Journal of the Mechanics and Physics of Solids*, vol. 67, pp. 129–151, 2014.

[MOR 05] MORTAIGNE B., "Vieillissement des composites. Mécanismes et méthodologie d'étude", *Techniques de l'Ingénieur*, 2005.

[MUL 73] MULLER I., *Thermodynamik*, Bertelsmann SE & Co., Düsseldorf, 1973.

[OLI 04] OLIVER W., PHARR W., "Measurement of hardness and elastic modulus by instrumented indentation: Advances in understanding and refinements to methodology", *Journal of Materials Research*, vol. 19, pp. 3–20, 2004.

[OLI 08] OLIVIER L., HO N., GRANDIDIER J.C. *et al.*, "Characterization by ultra micro-indentation of an oxidized epoxy polymer: Correlation with the predictions of a kinetic model of oxidation", *Polymer Degradation and Stability*, vol. 93, pp. 489–497, 2008.

[PAL 92] PALMESE G., Origin and influence of interphase material property gradients in thermosetting composites, PhD Thesis, University of Delaware, United States, 1991.

[PAP 02] PAPANICOLAOU G.C., MICHALOPOULOU M.V., ANIFANTIS N.K., "Thermal stresses in fibrous composites incorporating hybrid interphase regions", *Composites Science and Technology*, vol. 62, pp. 1881–1894, 2002.

[POC 08] POCHIRAJU K., TENDON G., SCHOEPPNER G., "Evolution of stress and deformations in high-temperature polymer matrix composites during thermo-oxidative aging", *Mechanics of Time-Dependent Materials*, vol. 12, pp. 45–68, 2008.

[PYT 12] PYTHON SOFTWARE FOUNDATION, Python, version 2.7.3, 2012.

[RAU 10] RAUCHS G., BARDON J., GEORGES D., "Identification of the material parameters of a viscous hyperelastic constitutive law from spherical indentation tests of rubber and validation by tensile tests", *Mechanics of Materials*, vol. 42, pp. 961–973, 2010.

[RIC 11] RICHAUD E., VERDU J., "Vieillissement chimique des polymères – Mécanismes de dégradation", *Techniques de l'Ingénieur*, 2011.

[SAN 09] SANDIA NATIONAL LABORATORIES, Dakota, version 5.2., 2009.

[SUB 96] SUBRAMANIAN S., LESKO J.J., REIFSNIDER K.L. *et al.*, "Characterization of the fiber-matrix interphase and its influence on mechanical properties of unidirectional composites", *Journal of Composite Material,* vol. 30, pp. 309–332, 1996.

[TRU 84] TRUESDELL C., *Rational Thermodynamics*, Springer-Verlag, Berlin, 1984.

[TVE 11] TVERGARD V., NEEDLEMAN A., "Polymer indentation: Numerical analysis and comparison with a spherical cavity model", *Journal of the Mechanics and Physics of Solids*, vol. 59, pp. 1669–1684, 2011.

[VER 12] VERDU J., *Oxidative Ageing of Polymers*, ISTE Ltd, London and John Wiley & Sons, New York, 2012.

[VU 12] VU D.Q., GIGLIOTTI M., LAFARIE-FRENOT M.C., "Experimental characterization of thermo-oxidation-induced shrinkage and damage in polymer matrix composites", *Composites Part A: Applied Science and Manufacturing*, vol. 43, pp. 577–586, 2012.

[YAN 03] YANG F., PITCHUMANI R., "A kinetics model for interphase formation in thermosetting matrix composites", *Journal of Applied Polymer Science*, vol. 89, pp. 3220–3236, 2003.

Index

Other titles from

iSTE

in

Materials Science

2017

CARVELLI Valter, JAIN Atul, LOMOV Stepan
Fatigue of Textile and Short Fiber Reinforced Composites
(Durability and Ageing of Organic Composite Materials Set – Volume 1)

DELHAES Pierre
Materials and Thermodynamics: Living and Economic Systems

FRESSENGEAS Claude
Mechanics of Dislocation Fields

2016

TORRENTI Jean-Michel
Materials and Infrastructures 1
Materials and Infrastructures 2

RÉMOND Yves, AHZI Saïd, BANIASSADI Majid, GARMESTANI Hamid
Applied RVE Reconstruction and Homogenization of Heterogeneous Materials

2015

JOUD Jean-Charles, BARTHÉS-LABROUSSE Marie-Geneviève
Physical Chemistry and Acid-Base Properties of Surfaces

LALEVÉE Jacques, FOUASSIER Jean-Pierre
Dyes and Chromophores in Polymer Science

2014

HAMAIDE Thierry, DETERRE Rémi, FELLER Jean-François
Environmental Impact of Polymers

LAFLEUR Pierre, VERGNES Bruno
Polymer Extrusion

2013

BATHIAS Claude
Fatigue Limit in Metals

LEXCELLENT Christian
Shape-memory Alloys Handbook

2012

DAVIM J. Paulo
Wear of Advanced Materials

DELHAES Pierre
Carbon Science and Technology: From Energy to Materials

GOUDEAU Philippe, GUINEBRETIÈRE René
X-Rays and Materials

SAANOUNI Khemais
Damage Mechanics in Metal Forming: Advanced Modeling and Numerical Simulation

VANBÉSIEN Olivier
Artificial Materials

VERDU Jacques
Oxidative Ageing of Polymers

2011

BATHIAS Claude, PINEAU André
Fatigue of Materials and Structures: Application to Damage

BATHIAS Claude, PINEAU André
Fatigue of Materials and Structures: Application to Design

PRIESTER Louisette
Grain Boundaries and Crystalline Plasticity

VIGNES Alain
Extractive Metallurgy 1: Basic Thermodynamics and Kinetics
Extractive Metallurgy 2: Metallurgical Reaction Processes
Extractive Metallurgy 3: Processing Operations and Routes

2010

BATHIAS Claude, PINEAU André
Fatigue of Materials and Structures: Fundamentals

CHATEIGNER Daniel
Combined Analysis

CHEVALIER Yvon, VINH TUONG Jean
Mechanical Characterization of Materials and Wave Dispersion V1
Mechanical Characterization of Materials and Wave Dispersion V2

DELHAES Pierre
Carbon-based Solids and Materials

2009

ALVAREZ-ARMAS Iris, DEGALLAIX-MOREUIL Suzanne
Duplex Stainless Steels

DAVIM J. Paulo
Machining Composite Materials

GALERIE Alain
Vapor Surface Treatments

2008

BATTAGLIA Jean-Luc
Heat Transfer in Materials Forming Processes

BLONDEAU Régis
Metallurgy and Mechanics of Welding

FRANÇOIS Dominique
Structural Components

REYNE Maurice
Plastic Forming Processes

TAKADOUM Jamal
Materials and Surface Engineering in Tribology

2007

BOCH Philippe, NIEPCE Jean-Claude
Ceramic Materials

CRISTESCU Constantin
Materials with Rheological Properties

2006

LASSEN Tom, RÉCHO Naman
Fatigue Life Analyses of Welded Structures

Printed and bound by CPI Group (UK) Ltd, Croydon, CR0 4YY